R. Droh R. Spintge (Eds.)

Innovations in Physiological Anaesthesia and Monitoring

With 98 figures

Springer-Verlag Berlin Heidelberg New York
London Paris Tokyo Hong Kong

Dr. med. R. Droh
Dr. med. R. Spintge

Sportkrankenhaus Hellersen
Paulmannshöher Straße 17
D-5880 Lüdenscheid, FRG

ISBN-13: 978-3-540-50331-6 e-ISBN-13: 978-3-642-74114-2
DOI: 10.1007/978-3-642-74114-2

Library of Congress Cataloging-in-Publication Data
Innovations in physiological anaesthesia and monitoring / R. Droh, R. Spintge (eds.) p. cm.
ISBN-13: 978-3-540-50331-6 (U.S.: alk. paper)
1. Anesthesia. 2. Patient monitoring. I. Droh, R. (Roland)
II. Spintge, R. [DNLM: 1. Anesthesia. Monitoring, Physiologic. WO 200 I58] RD82.I53
1989 617.9'6--dc20 DNLM/DLC for Library of Congress 89-21627

2119/3140-543210 – Printed on acid-free paper

Preface

At present there are going on decisive developments in the introduction of physiological thinking into anaesthesiological practice. These innovative motion comprises applications taken from space physiology, closed-circuit ventilation systems, neurophysiologic control of circulation and ventilation together with innovative aspects of circulation physiology and respiration physiology. Involved are innovative aspects of neuromuscular blockade, volatile anaesthetics, cardiorespiratoric monitoring, haemodilution and blood substitution.

Special emphasis is put on demonstrating the closed circuit system as a combination of anaesthesia system and monitoring system in daily clinical routine. Additional knowledge and investigation are stimulated in the fields of anaesthesia, neurophysiology, pharmacology, space medicine and metabolic physiology using non-invasive methods of monitoring like the evaluation or the neurovegetative status of man and its addaptive changes. Combining these aspects and systems will provide a quite new analytic and synergetic power in the direction of physiological anaesthesia and monitoring which finally will result in a totally new system of a physiological "micro-anaesthesia".

We recommend this bock therefore to all anaesthetists, clinical physiologists and pharmacologists as well as to physicists and engineers interested in innovative and non-invasive methods in physiological anaesthesia with special emphasis to the use of closed systems.

The editors

Contents

List of Contributors

H.-H. Abel, Institut für Physiologie der Freien Universität Berlin, Arnimallee 22, 1000 Berlin (West) 33, Germany

P. C. W. Beatty, Regional Department of Medical Physics and Bioengineering, University Hospital of South Manchester, Withington, Manchester M20 8LR, UK

S. M. Cain, Department of Physiology and Biophysics, University of Alabama at Birmingham, Alabama 35294, USA

A. W. Cuiper, Institute of Anaesthesiology, Oostersingel 59, 9713-EZ Groningen, The Netherlands

W. L. den Dunnen, Department of Anaesthesia, Municipal Hospital Dordrecht, Dordrecht, The Netherlands

J. C. Dorlas, Instituut voor Anaesthesiologie, Academisch Ziekenhuis Groningen, 9700 RB Groningen, The Netherlands

R.Droh, Sportkrankenhaus Hellersen, Paulmannshöher Str. 17, 5880 Lüdenscheid, FRG

W. Erdmann, Department of Anaesthesiology, Erasmus University, Academisch Ziekenhuis, Molewaterplein 40, 3015-GD Rotterdam, The Netherlands

N. S. Faithfull, Department of Anaesthesia, University of Manchester, Manchester, UK

M. Fennema, Department of Anaesthesiology, Erasmus University, Academisch Ziekenhuis, Molewaterplein 40, 3015-GD Rotterdam, The Netherlands

T. E. J. Healy, Department of Anaesthesia, University Hospital of South Manchester, Withington, Manchester M20 8LR, UK

P. Heinrich, Institut für Physikalische und Theoretische Chemie, Universität Essen, 4300 Essen 1, FRG

W. Ischebeck, Neurochirurg. Abteilung und Institut für Kernspintomographie, Gemeinschaftskrankenhaus Witten/Herdecke, 5804 Witten/Herdecke, FRG

B. Kay, University Department of Anaesthesia, University Hospital of South Manchester, Withington, Manchester M20 8LR, UK

H. P. Keller, Universitätsklinik Würzburg, 8700 Würzburg, FRG

K. A. Kirsch, Institut für Physiologie der Freien Universität Berlin, Arnimallee 22, 1000 Berlin (West) 33, Germany

D. Klüßendorf, Institut für Physiologie der Freien Universität Berlin, Arnimallee 22, 1000 Berlin (West) 33, Germany

H.-P. Koepchen, Institut für Physiologie der Freien Universität Berlin, Arnimallee 22, 1000 Berlin (West) 33, Germany

H.-D. Kuntz, Abteilung für Gasteroenterologie und Hepatologie, Medizinische Universitäts-Kliniken, Bergmannsheil Bochum, 4630 Bochum, FRG

H. Lauboeck, Zentrale Anaesthesieabteilung, Bergmannsheil Bochum, 4630 Bochum, FRG

D. W. Lübbers, Max-Planck-Institut für Systemphysiologie, Rheinlanddamm 201, 4600 Dortmund, FRG

A. Matsuki, Department of Anaesthesiology, Hirosaki University School of Medicine, Hirosaki, Aomori-Ken, Japan

U. Merschebrock, Max-Planck-Institut für Systemphysiologie, Rheinlanddamm 201, 4600 Dortmund, FRG

T. Mostert, Department of Anaesthesia, Municipal Hospital, Dordrecht, The Netherlands

T. Murakawa, Department of Anaesthesiology, Hirosaki University School of Medicine, Hirosaki, Aomori-Ken, Japan

T. Oyama, Department of Anaesthesiology, Hirosaki University School of Medicine, Hirosaki, Aomori-Ken, Japan

L. Röcker, Institut für Physiologie der Freien Universität Berlin, Arnimallee 22, 1000 Berlin (West) 33, Germany

Y. Satoh, Department of Anaesthesiology, Hirosaki University School of Medicine, Hirosaki, Aomori-Ken, Japan

B. Schrader, Institut für Physikalische und Theoretische Chemie, Universität Essen, 4300 Essen 1, FRG

W. Schregel, Abteilung für Anaesthesie und operative Intensivtherapie, Knappschaftskrankenhaus Bochum-Langendreer, Universitäts-Klinik, In der Schornau 23–25, 4630 Bochum, FRG

M. Vitt, Abteilung für Anaesthesie und operative Intensivtherapie, Knappschaftskrankenhaus-Universitäts-Klinik, In der Schornau 23–25, 4630 Bochum, FRG

R. Wyzgol, Institut für Physikalische und Theoretische Chemie, Universität Essen, 4300 Essen 1, FRG

Terrestrial And Extraterrestrial Applied Physiology

Physiological Concepts of Cardiovascular and Respiratory Control: Theoretical Basis and Applicability in Man

H.P. KOEPCHEN, H.-H. ABEL and D. KLÜSSENDORF

Introduction

Key concepts of the present symposium are: anaesthesia – physiology – internal medicine – innovation. What are the relationships among these concepts? We believe one of the innovations to lie precisely in their combination. Internal medicine, and especially anaesthesia, is to a considerable extent applied physiology. The still too rigid borders between the disciplines, especially in Germany, must be overcome to make the concepts and methods developed in one field fruitful for application in other fields. This is especially important for the application of basic physiological concepts to man. In this context we will confine our attention to one central point of view, that of neurovegetative regulation, in particular of the cardiovascular and respiratory systems. The term neurovegetative means neuronal regulation of the functions of vegetative organs and organ systems. Within this general framework the cardiovascular and respiratory regulatory systems form part of an interacting complex of many kinds of regulatory processes.

The special importance of cardiovascular and respiratory regulation in this context is based on several facts. Circulation and respiration are basic, vital functions of the body. Their adequate maintenance is one of the most important aims of the anaesthesiologist. In this area there is therefore a convergence of the interests of physiology and those of practical medicine. Both functions are dependent on central control systems which are intimately linked with the functional state of general brain function. This is true even during rest and anaesthesia. In addition, these control systems are influenced by higher cortical functions during wakefulness, mental load and exertion, i.e. during the different degrees of human activity. The consequence is that the behaviour of respiratory and cardiovascular parameters mirrors the state of central nervous regulations and can be used as an indicator of them.

Modern techniques allow one (a) to record these parameters continuously without harm to the patient and (b) to evaluate automatically correlations between single parameters so as to reveal their time structures and to process a great amount of data within a short time.

Much information on the functions of vital control systems is thus available which could be used for the characterization of the neurovegetative state in humans. This applies to the clinical patient as well as to healthy subjects in normal, everyday conditions and their adaptive changes, e.g. physical training, adaptation to an unusual environment such as microgravity, or rehabilitation after diseases or surgical interventions. In other words, nature provides us with signals which we do not use sufficiently. This makes innovation in this field not only possible but urgently necessary, since (a) we are committed to the best possible care of patients and (b) similar trends are developing rapidly in other countries where they are influencing clinical medicine.

There is, however, one danger. The unreflected collecting of easily accessible cardiorespiratory (i.e. cardiovascular-respiratory) data and their feeding into computers would provide us with a huge amount of information which we must learn to interpret – we must be able to discriminate between important information and "noise". Otherwise, our situation could come to resemble that of the greedy man in the fable who was killed by the rain of ten thousand ducats which he had asked his magic ring to give him at once.

At this point we need to consider the concepts which physiology has developed over recent decades. Here we must be careful however, since in the present era of sophisticated techniques and studies of single elements the theories of whole systems have not been followed up in an adequate manner. Many reviews in textbooks, and especially the compressed physiological introductions in textbooks of internal medicine or anaesthesia, are indebted to classical physiological concepts which must be critically reexamined.

Therefore, before coming to the application in man, we will present an overview of the main physiological concepts in cardiovascular and respiratory nervous regulation and the corrections which must be made in them today. We will begin with cardiovascular regulation, continue with respiratory regulation, and then come to the interrelations between them. Furthermore, we will consider the methods available for measurements in man, their functional meaning and interpretation and will present initial results in this direction from our own group; these are outlined more extensively in the chapter by Abel et al. (this volume). Possible further developments will then be considered.

Survey of Common Physiological Concepts

Classical Physiological Concepts of Cardiovascular Regulation [16]

Since the discovery of the aortic baroreceptors by Cyon and Ludwig in 1866 and the carotid sinus baroreceptors by Hering in 1927, baroreceptor reflexes have been in the center of studies on and concepts of cardiovascular control physiology. The term "reflex control of circulation" [13] is often used for the entirety of cardiovascular control. This concept of reflex control of arterial pressure was elaborated in a quantitative manner in recent decades by introducing modern control theory and treating the whole system as a closed feedback loop, taking into account the feedback action of the regulated value on the receptors [12, 34, 39]. In our opinion, this theoretical approach had no important consequences for clinical application or for the understanding of pathophysiological deviations. The basic idea is the following: A system which, by simultaneously controlling the inflow to and the outflow from the arterial system, maintains a constant level of arterial blood pressure provides all organs with a sufficient source of driving pressure. The demand of different organs then can be met by peripheral, local regulation. This concept may be basically correct, but needs many modifications. One of these is the maintenance of an adequate filling by the control of water intake and excretion; the importance of the low-pressure system was outlined by Gauer and his school [9]. Another modification is the fine microcirculatory control of tissue perfusion, where Weigelt and Lübbers [40] have recently given new impulses. Nevertheless, the high-pressure system control has the advantage of occuring in the order of tens of seconds and is thus suitable for monitoring over a short period in which arterial blood pressure is an easily accessible parameter, integrating the whole cardiovascular system. The problem is that blood pressure is not an unequivocal indicator for the most vital circulatory function, i.e. providing an adequate blood supply to the organs. Recently, the continuous recording of arterial blood pressure gained renewed interest and application as the stimulus for the baroreceptor reflex control, and its use was extended beyond the routine monitoring during surgery and intensive care. We will return to this point below.

Classical Physiological Concepts of Respiratory Regulation [17]

Physiological concepts of respiratory regulation are centered around the main function of respiration, i.e. the exchange of O_2 and CO_2 between the body and the environment. Here there are two branches of research, and these are usually performed in separate laboratories. The first deals with adaptation of the respiratory motor act to the changing demands of gas exchange in terms of the variable intensity of metabolism. This concerns both, the respiratory drive by CO_2 and hypoxia and the central and peripheral receptors mediating these drives. In many modern textbooks CO_2 is still considered the special drive for respiration, whereas other influences vary the sensitivity against CO_2. In analogy to the blood pressure control circuit, the CO_2 feedback system can be seen as a closed circuit and described with the methods and nomenclature of mathematical feedback control theory [28]. In this respect the respiratory drive P_{CO_2} is the controlled variable, and the purpose of the circuit is to hold P_{CO_2} constant. This concept approaches reality the more the subject or animal approaches a reduced state of activity, e.g. with anaesthesia or neuromuscular blockade, where in contrast to the circulatory feedback the control circuit is opened and the central command taken over by the anaesthesiologist. It would be a step forward if all physicians who practise this regime were to realize the classical function of the feedback circuit and the central importance of CO_2 as the controlled variable.

Under special conditions, a second feedback circuit comes into play in the form of the hypoxic drive via peripheral chemoreceptors, which in a fairly complicated manner interferes peripherally and centrally with the CO_2 feedback system. Here, again, respiratory control is under the dominant influence of the feedback system, especially during strongly reduced states of the organs. Modern methods – an essential part of which have been developed in the Max-Planck-Institut in Dortmund under the guidance of Lübbers – also allow reliable quantitative monitoring ot the P_{O_2}.

The second branch of respiratory control physiology exerted mainly by neurophysiologists is directed to the central origin and mechanism of respiratory rhythmogenesis. Respiratory rhythmicity is the prerequisite of moving air in and out of the thorax, and as a "regular" function it has therefore been investigated thoroughly – in contrast to cardiovascular rhythmicity, which is often considered as something of a curiosity. Recently, increasingly sophisticated neurophysiological theories on the mechanism of respiratory rhythmogenesis have been developed, mainly in the laboratories of von Euler [8] in Stockholm, Cohen [6] in New York and Richter [35] in Heidelberg. These researchers claim to be able to explain completely the mechanism of rhythmogenesis by, in part, fairly deterministic models. It is not necessary in the present context to go into the details of these theories and the ongoing controversies about them, but some crucial points should be mentioned which, on the one hand, can have pathophysiological implications and, on the other, provide a basis for the analysis of respiration in man. There is now nearly general agreement that the motor act of respiration is governed by several different components which independently control the various parts of the respiratory cycle. On the other hand it seems proven that the respiratory drive, especially the chemical drive within the feedback system, acts by way of structures which are not identical with the rhythmogenic apparatus. The latter recognition is based mainly on the work of Loeschcke [29], Loeschcke and Koepchen [30] and Schlaefke [36].

One example of the contributions of our own laboratory to understanding the mechanism of respiratory rhythmogenesis in animal experiments illustrates further relationships to the application in man. In anaesthetized cats it was possible to find a small area in the brainstem in the vicinity of the solitary tract, the lesion or local cooling of which leads to an enormous prolongation of inspiratory duration after exclusion of vagal pulmonary stretch afferents (Fig. 1) [20, 25]. It is important that this experimental disturbance does not change the expiratory duration, i.e. the starting mechanism of the next inspiration, and likewise leaves intact the effectiveness of the chemical drive, which is expressed, for example, in the steepness of increase of phrenic bursts controlling the velocity of inspiratory air flow. This experiment demonstrates (a) that the termination of inspiration is an active process and (b) that it can be influenced separately from other partial components of respiratory rhythmogenesis. The partial component governing inspiratory velocity ("ramp generator" according to von Euler [8]) together with the counteracting component switching off the inspiratory phase determine the tidal volume. As Abel et al. demonstrate in the companion presentation, it can be shown in man that also here tidal volume is the resulting and not the primary parameter. This opens

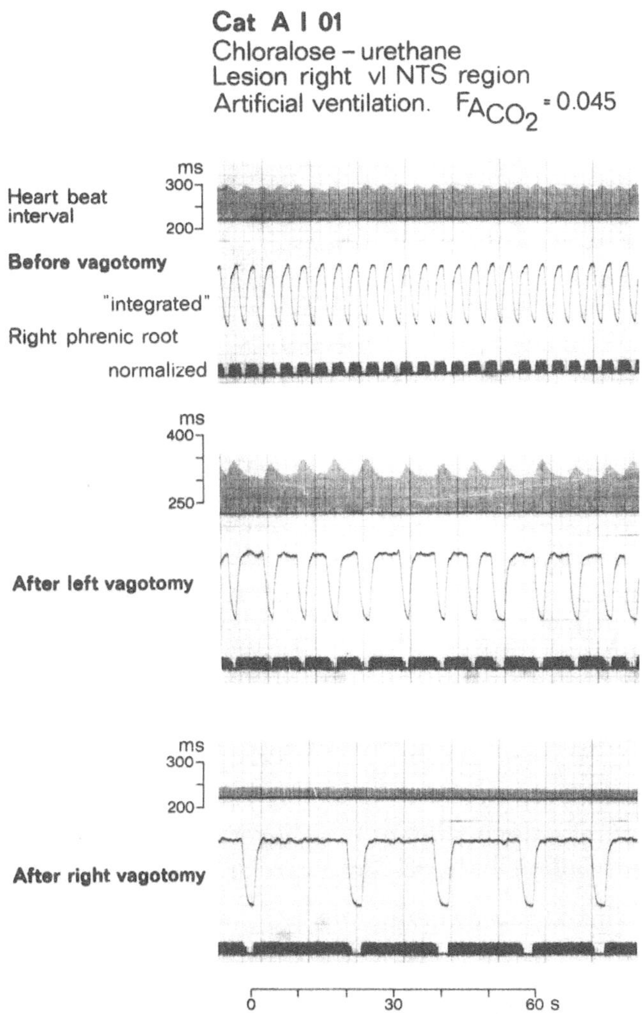

Fig. 1. Isolated impairment of the respiratory partial component of termination of inspiration by a small local lesion in the region ventrolateral to the solitary tract in an anaesthetized, immobilized, artificially ventilated cat. *Upper trace,* heart beat interval; *middle trace,* electrically "integrated" activity of a phrenic nerve rootlet; *lower trace,* normalized phrenic discharges. In the *top record* vagi are intact. The effect of the lesion becomes manifest only after vagotomy. Prolongation of phrenic discharge after unilateral vagotomy *(middle record)* is extremely enhanced after bilateral vagotomy *(bottom record).* The lesion does not essentially change other partial components of respiratory rhythmogenesis as rate of rise of increase of phrenic discharge ("ramp"), amplitude of maximum phrenic discharge or expiratory time. (From experiments of H.-P. Koepchen, H. Lazar, D. Klüßendorf and T. Hukuhara)

the possibility to classify respiratory disturbances in man according to the partial components which are involved and to separate a deficiency of drive from disturbances of rhythmogenesis.

Classical Concepts of Interrelations Between Cardiovascular and Respiratory Control [14, 19, 22]

Before entering into a critical examination of the outlined physiological concepts of circulatory and respiratory control, we will deal with the interrelations between these control systems, because it is precisely the study of the interrelations that has led to important modifications of the theoretical picture. From the functional point of view, a close connection between the two control systems lying in series with respect to gas exchange is an inevitable postulate. But the mechanisms are still unclear regarding the basic question as to whether the coupling is due to intracentral interconnections or to peripheral reflex afferents influencing both control systems. The first, seemingly self-evident supposition is that there are two separate control systems, one for respiration and the other for circulation. This postulate stems largely from the fact that research on respiratory control and that on circulatory control generally take place in different laboratories. This separation also favours the conceptual constructions of specific respiratory or cardiovascular reflexes, with the corresponding afferences and reflex centres. Since, on the other hand, findings as to the action of cardiovascular afferents on respiration and vice versa as well as respiratory rhythmicity in the cardiovascular innervation are unquestionable, the terms "side effect"and "irradiation" were suggested as a way to maintain the original presumption of separate control systems. These terms obscure the fact, however, that our decision as to what is a specific function and what is a side effect remains purely arbitrary and based on teleological anticipation.

Critical Examination of the Classical Physiological Concepts of Correlations Between Cardiovascular and Respiratory Control

The problems involved in the conventional concepts become evident when we consider in an unbiased way the interrelations between cardiovascular and respiratory control. The arguments here relate to two kinds of experimental findings: (a) the observation of peripheral rhythmic events and (b) the neurophysiology of brainstem control systems. Only those findings are selected here from the wealth of results which are relevant for our basic theme, i.e. the applicability to man.

Observation of Peripheral Rhythmic Events

Every continuous recording demonstrates that cardiovascular innervation – cardiac as well as vasomotor – is not tonic but basically rhythmic. Both explanations which try to avoid acknowledging primary cardiovascular rhythms are untenable; one declares them purely a side effect of an unstable oscillatory baroreceptor feedback circuit, the other as an irradiation from the respiratory centre. First, oscillations continue when the baroreceptors are denervated. Secondly, many observations show that cardiovascular rhythms are not necessarily secondary to a fixed dominating respiratory rhythm, but that the two

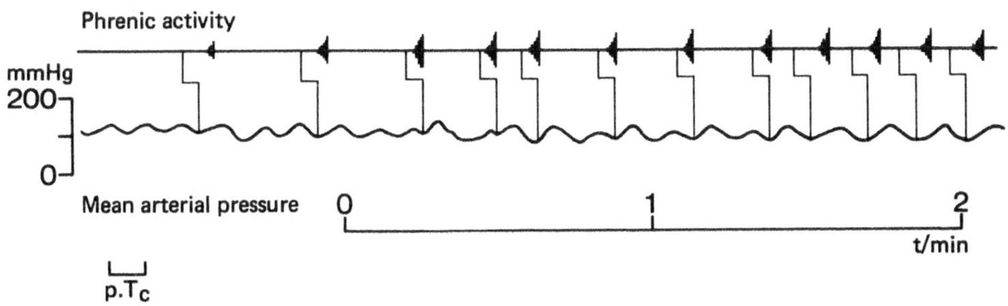

Fig. 2. Various kinds of relative coordination between respiratory and blood pressure rhythms in an anaesthetized, immobilized dog under diffusion respiration. *Upper trace,* activity in a root of the phrenic nerve; *lower trace,* mean blood pressure. Immediately after the injection of succinylcholine (before this recording) phrenic activity ceased temporarily, whereas blood pressure fluctuations in the rhythm of the former respiratory rhythm continued. During the reappearance and gradual enhancement of phrenic activity various kinds of shifting coordination between the respiratory and the cardiovascular rhythm can be seen. These lead through a state of temporary 2:1 coordination to synchrony between the two rhythms. This example shows clearly that synchrony is the result of shifting mutual influence and not of "irradiation" from the respiratory to the cardiovascular control system. [From 14]

rhythms influence each other mutually. Two examples from experiments in anaesthetized, curarized, artificially ventilated dogs illustrate this (Fig. 2). Here, after an injection of succinylcholine both respiration and central respiratory innervation stopped for several minutes, manifested by the disappearance of phrenic bursts, while blood pressure variations continued with the frequency of the foregoing respiratory rhythm. This demonstrates, firstly, that the previous cardiovascular rhythm was synchronized with, but not caused by, the respiratory rhythm. The process of synchronization is quite obvious in the subsequent course of the recording, when an irregular phrenic rhythm slowly reappears and now interacts with the blood pressure rhythm in such a way that they influence each other, resulting initially in a 2:1 coordination and, finally, in a synchronization through changes in both frequencies.

Another instructive example shows the time course of blood pressure fluctuations and phrenic nerve activity during high-frequency hyperventilation, likewise in a curarized anaesthetized dog (Fig. 3). One can observe a first phase during which phrenic bursts and blood pressure waves are synchronized. According to the classical, and still common, interpretation these would be "respiratory" blood pressure waves caused by the respiratory centre as an "irradiation". In the further course of the experiment the phrenic bursts disappear, but the blood pressure rhythm continues, as clear evidence that previously the latter was not secondary to respiratory rhythm, but that both had been synchronized. Golenhofen and Hildebrandt [10] showed similiar shifting correlations between respiration and blood pressure oscillations in man, where in contrast to the animal experiment peripheral feedback cannot be excluded. Our animal experiments, however, strongly favour their hypothesis that also in man these coordinations are mainly due to central interactions. The reason that we return to these findings of over 20 years ago is that the same principles are

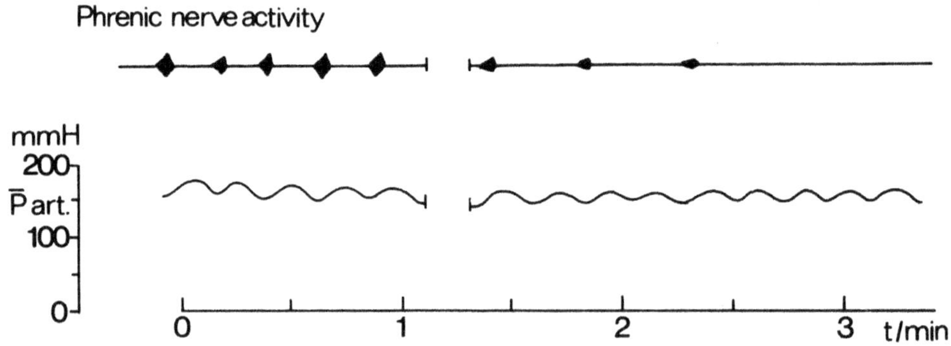

Fig. 3. Various kinds of coordination between respiratory rhythm recorded from a phrenic nerve root *(upper trace)* and oscillation of mean arterial blood pressure *(lower trace)* in an anaesthetized, immobilized dog under diffusion respiration *(left)* and subsequent artificial high-frequency hyperventilation *(right)*. Note disappearance of phrenic activity under hyperventilation after a period of 2:1 coordination, but continuation of blood pressure oscillations. This example demonstrates that the "respiratory" blood pressure waves at the beginning were the result of synchronization between both rhythms and not "irradiation" from the respiratory to the cardiovascular control system. [14]

being rediscovered in the modern computer analysis of cyclical cardiovascular events in humans without realizing the true interpretation of these phenomena, since the classical errors in interpretation have been repeated. We will return to this point below.

The essential result of the analysis of rhythm coordination in animal experiments can be summarized as follows: There exists centrally a neuronal network which generates rhythms that synchronize and desynchronize and can come into mutual resonance. The different efferent outputs are not strictly bound to special rhythms, so that, for example, the cardiovascular outputs can participate in the same rhythm as the respiratory innervation, but the latter can likewise be entrained by slower rhythms. Synchronized states and states with integer multiple relations are especially stable. Concerning the special phenomenon of blood pressure waves and correlated heart rate fluctuations in the range of 2–4 per minute, the result of long controversial discussions can be illustrated by the scheme of Fig. 4. This scheme illustrates that there are at least three interfering oscillatory systems: (a) the primary centrogenic neuronal rhythm, or, better, complex of rhythms; (b) the baroreceptor feedback circuit, which can oscillate because it comprises the slowly contracting vascular smooth muscles; and (c) the vascular smooth muscle itself which even without phasic innervation is a rhythmogenic system. The interference between rhythmic vasomotor innervation patterns and the rhythmic response of vasomotricity was shown in earlier studies [38] in which artificial nervous rhythms elicited resonance-like phenomena. These results from animal experiments should be kept in mind when recording rhythmic vasomotor events in the periphery in humans. These are often simply taken as indicators of the central nervous sympathetic innervation pattern without regard to the contribution of the time structure of the vascular muscle itself.

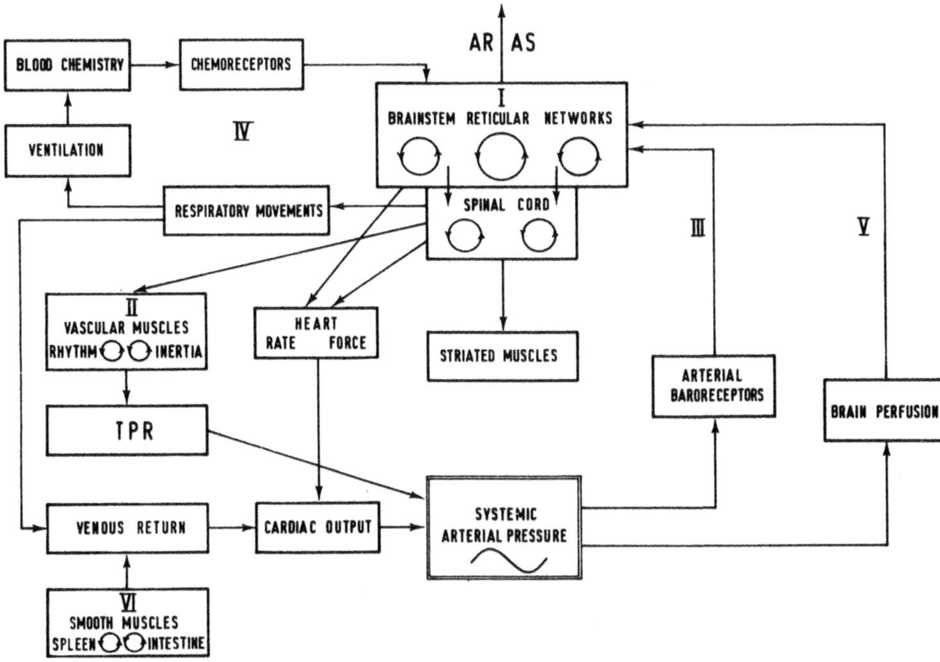

Fig. 4. Functional scheme showing the interference of several oscillatory systems in the generation of arterial blood pressure fluctuations. A synchronizing and desynchronizing complex of brainstem rhythms *(I)* provides heart and blood vessels with rhythmic innervation. Vasomotor rhythmic innervation interferes with the endogenous rhythmicity of the vascular smooth muscle *(II)*, resulting in oscillations of peripheral resistance. The oscillating arterial blood pressure produces rhythmic afferent activity in the baroreceptor nerves *(III)* fed back into the central nervous system. Under special conditions the respiratory feedback control system with its peripheral and central chemosensitivity *(IV)* is likewise capable of developing oscillations. In pathological cases, insufficient brain blood flow can induce increase in arterial pressure (Cushing reflex) which in turn improves brain perfusion, thus forming another oscillatory feedback circuit *(V)*. [From 19]

Neurophysiology of Brainstem Control Systems

The second basic criticism of the classical concepts of interrelation between two separate control systems comes from microrecordings in the brainstem performed in order to study the properties of the central substrate of the control systems. Of course, the prerequisite for this is identification of the neurones belonging to the respective control systems. Accordingly, many authors have labelled central neurones as "respiratory" or "cardiovascular" or "vasomotor". Usual criteria are the relation to the phrenic rhythm for respiratory neurones and the response to baroreceptor stimulation or the correlation to spontaneous blood pressure changes for cardiovascular neurones. This classification neglects the fact that such criteria are necessary but not sufficient [21] since it is known that baroreceptors do not act only on cardiovascular efferents, that blood pressure variations are paralleled by other phenomena, and that respiratory-related rhythm is present in many nonrespiratory outputs of the nervous system. Without pursuing this argument further, the fact becomes evident that the central control substrates are at least partly identical [18,23]. Thus the picture of a multifunctional neuronal network emerges which can change its relative contributions to different outputs. In such a network specificity is a quantitative and not a qualitative property of any single element. The common central neuronal network is at the same time the substrate for the afore-mentioned common central rhythmicity. This theory at the same time solves the problem of the specific cardiovascular or respiratory or other reflexes, including the now dispensable concepts of "irradiation" and "side effects". An arbitrary model network illustrates this kind of gradually specific common network (Fig. 5). This model shows that a common network does not at all produce identical patterns in all outputs, but that even a slight nonhomogeneity within the network results in an extremely uneven distribution of the input-output relations; this means a certain degree of specificity, as can be observed in the natural set of outputs coming from cardiovascular and respiratory control substrates.

This model provides a tool for analysing the network from both directions. From the output side we can ask: how much do the different inputs contribute to this special output? From the input side, on the other hand, we can ask: how much does a special input express its activity in the various outputs? We must realize that this network comprises the different rhythms which can be used as markers; here the additional questions can be asked as to what the action of a special input to a special rhythm might be, and what rhythm appears with what intensity in the output under study. At first this approach seems to be quite theoretical, but it can be translated into very concrete questions for investigations even in man, for example: in what manner are various parameters, e.g. heart rate, vasomotor innervation or respiratory pattern, and their rhythmicities, influenced when we change the input to central chemoreceptors, i.e. when we elevate the alveolar P_{CO_2}?

There is another consequence of the common network model, especially with respect to investigations in intact man. The partly common substrates comprise not only sources of respiratory and cardiovascular innervation but also the brainstem structures which, according to the work of Magoun and Moruzzi [31], are responsible for the control of muscular tone [15, 26]. This makes understandable the strong correlation between actual sympathetic tone and ventilatory drive and the state of alertness and mental activity, a point which is scarcely accessible in animal experiments.

Application of the Physiological Background to Man

After this critical examination and revision of the physiological background we can now consider the applicability to man. Here we must say in advance that the question as to how far these results – obtained mainly in animal experiments and printed under the heading "Human Physiology" in textbooks – can be translated to man, is one of the topics of our present research. Certain differences and pecularities of the studies in humans can nevertheless already be noted and should be kept in mind as permanent presuppositions.

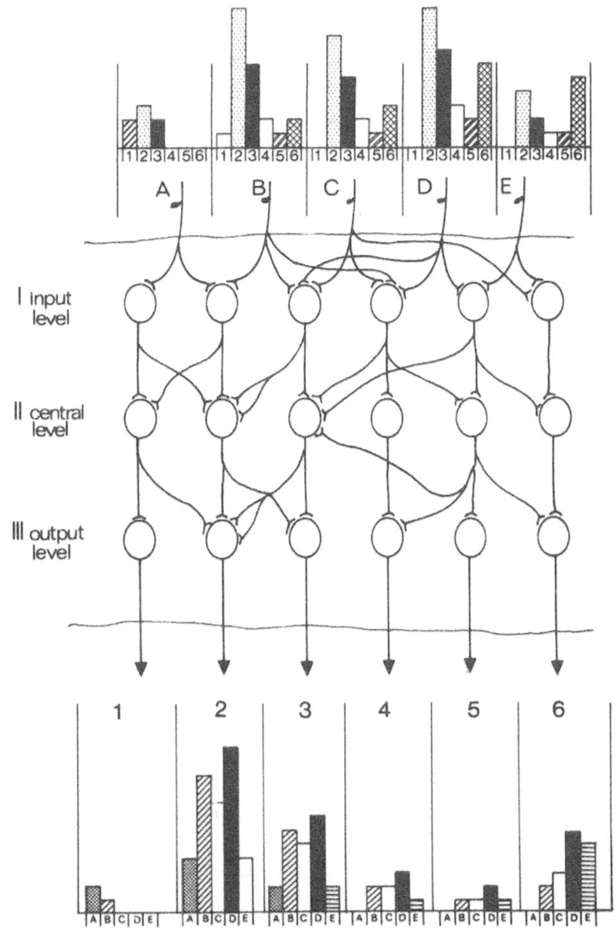

Fig. 5. Theoretical model of a central neuronal network common for several input-output relations. The model was calculated under the extremely simplified assumption of equal excitation in all inputs and pure additive excitatory synaptic processing. *Columns at top:* view from the input side into the network; relative amount of activity in the different outputs coming from the respective input. *Columns at bottom:* view from the output side into the network; relative amount of activity in the respective output coming from the different inputs. The model illustrates that in a common network slight nonhomogeneities of synaptic connections can lead to considerable inequality in the input-output relations, i.e. graded specificity. It must be assumed that the distribution of activity within the network is not static but changes with the changing relation of activity in the various inputs, with rhythms within the network and additional local inhibitions. [From 18]

Relative Importance of Feedback Reflex Control

We have seen that the reflex approach, further developed by the closed circuit control theory, governs the physiological concepts of the homeostatic autoregulatory systems. Here the reflex is the primary event derived from the artificial manner in which physiologists test the function of the control systems. One example is the recently renewed interest in the baroreceptor reflex as applied in man, for example, by Litter et al. [27] in England and Mancia et al. [33] in Italy, both in clinical patients. Another example is the investigation of the CO_2 sensitivity of respiration. In this approach, central influences are seen as disturbances, or "noise", coming from higher structures and interfering with the feedback control. They can be described as changes in reflex sensitivity. This approach is justified and was successful in the anaesthetized animal, where the higher influences have been removed by anaesthesia and the basic control circuits work in a reproducible manner particularly, when the system is "normalized" by exclusion of the so-called disturbances from higher structures.

We must realize, however, that in the normally behaving animal, and even much more in the unrestrained human being, the behavioural influences from above are more important than the feedback circuits. Consequently the changes of parameters that we see are much more the expression of these behavioural factors than of the homeostatic feedback systems which, of course, are not out of function but only prohibit overly great deviations from the physiological range. Thus, for example, when we observe the relation between alveolar P_{CO_2} and ventilation in the healthy alert human, ventilation does not follow the P_{CO_2}, but P_{CO_2} goes secondarily in the reverse direction (for an example see Abel et al., this volume).

We must realize that this change between the preponderance of behavioural factors and that of the automatic feedback control takes place when a person changes from a healthy state of active behaviour to that of a resting, restrained patient and even more to anaesthesia or the conditions of intensive care, where he approaches the basic state of the experimental animal. It is a striking challenge for the clinician as well as for the physiologist to develop methods to characterize and monitor this transition.

Spontaneity, Individuality and Vegetative Innervation

Spontaneity and individuality are basic human properties. Classical neurophyiological research was oriented very much by the reflex approach; the origin of voluntary movements has thus remained a mystery up to the present, and spontaneity in physiological concepts still plays the role as a disturbance for a "proper" regular reflex study. What does this have to do with the recording of cardiorespiratory and other vegetative parameters? In the common physiological concepts, the adaptive changes of vegetative innervation at physical exertion are considered secondary to motor activity, caused by irradiation from the motor cortex to vegetative brainstem centers and by feedback drives from working muscles. In contrast, Heinemann et al. [11] were the first to show that with activations of limbic structures, which also in humans are believed to be responsible for the motivation of motor behaviour, vegetative innervation does not follow, but precedes the movement. Thus in the awake human the play of motivations, even if it is still subthreshold for motor innervation, is expressed continuously in vegetative parameters and can be detected with modern recording methods.

This fine play, of course, is highly dependent on the individual personality. In the usual scientific approach this is expressed in a great variability which makes it difficult to reach statistically significant results. In the research on man, however, especially in patients, one should keep in mind that individuality is at least as important from the diagnostic and therapeutic points of view as it is in the statistical study of the behaviour of large groups. Thus, one of the principles in the analysis of vegetative patterns is the consideration of both these points of view. The practical consequence of this is that a great number of subjects must be tested to establish a mean reaction while at the same time looking for the individual pattern in relation to the mean. This principle can be extended to specific groups, e.g. trained sportsmen or patients suffering from particular diseases.

Role of Rhythmicity in Research on Humans

Rhythmicity can be considered an intermediate state between the nonsequential spontaneity of events and the nonvariable rigidity of parameters in reduced states in which rhythmicity is reduced or lost, while the basic feedback regulations are still in function. Rhythmicity is the physiological state during rest in which vegetative spontaneity is characterized by certain structures over time, with preferential but not fixed ranges of frequency. The analysis of these can therefore help to gain quantifiable characterization of the acutal state of regulation and activity. It can be shown that rhythmicity is a parameter which can change independently of changes in mean values, reacts more sensitively, and outlasts the deviation of the mean value after different kinds of mental or physical load.

Interference Between Measurements and Results

It is a basic principle in scietific research that the measurement itself should not influence the results. The more stable and resistant are the parameters against interferences by the procedure of their measurement, the more radical can be the invasion by the measurement. This applies, for instance, to the basic state of deep anaesthesia in animal experiments and in man. In contrast, the fine play of innervation patterns in the awake human is extremely sensitive to any kind of experimental invasion. For example, the respiratory mask or mouthpiece evokes a profound change in respiratory pattern. The same applies to any other kind of instrumental invasion. Therefore, we follow strictly the principle of noninvasive methodology, even if this results in a reduction of obtainable information. The situation changes, of course, when monitoring by invasive methods is practised routinely during surgery and intensive care. Here the extended possibility to obtain data should be used, keeping in mind the changed conditions.

Remarks on our Initial Results [1–2, 24]

We must note, at the outset, that we are presently far from completing the theoretical and conceptual frame outlined above. Our initial investigations have not been fully guided by a systematic approach and using a definite concept but have been influenced by the application of available methods and the still limited possibilities of computer evaluation. The accompanying presentation (Abel et al.) demonstrates the first, individual islands, as it were, in the great sea of unexhausted possibilities of neurovegetative research in man. We will conclude, therefore, with some hints as to where these findings are to be placed within the overall concept.

Heart Rate

According to the question posed, we can measure mean heart rate, or heart rate variability, or, in order to specify the frequency ranges of heart rate rhythmicity, we can compute power spectra of heart beat invervals. The latter method confirms that there are three main ranges of frequency. One is in the order of normal respiratory frequency, the second is in a lower frequency range of a few periods per minute, and, finally, there is a very low frequency range in the order of a duration of minutes, which we have not yet considered in our evaluations. Here we are basically in agreement with other authors [32], especially Eckoldt et al. [7] in East Berlin, who have been performing heart frequency analyses in man for several years. Referring to the rhythm analysis in anaesthetized dogs mentioned above, the frequency ranges of heart rate reported correspond to the preferential frequency of respiration and the lower preferential frequency of blood pressure waves in animals as well as in man. Accordingly, they are frequently called "respiratory" and "blood pressure" frequency ranges, respectively. This is related to the conventional discrimination between respiratory and so-called Mayer waves. We avoid this nomenclature because animal experiments have shown that it is misleading insofar as respiration is not a fixed reference but can move out of the higher frequency range without disappearance of the corresponding cardiovascular rhythmicity. This means that in man, too, cardiovascular rhythmicity in the frequency range of respiration is not a side effect of the respiratory rhythm. As one indicator for this fact, we found that the tidal volume is not strongly related to the amount of heart rate variability in this range.

On the other hand, in agreement with the model of multiple synchronizing and desynchronizing oscillators, respiratory frequency can shift into the lower range and synchronize with the rhythms in that range. This is accompanied by resonance-like enhancement of heart rate fluctuations. This phenomenon was described by Angelone and Coulter [5] during voluntary slowed respiration. We perform this experiment routinely with medical students, but in our research work we do not apply this prescribed respiration since it introduces an additional mental load which can change the conditions. In more recent evaluations, however, we have seen that the same phenomenon occurs spontaneously. This has the consequence that the actual length of the respiratory cycle must be regarded when we judge the amount of sinus arrhythmia [3]. Since sinus arrhythmia is already correlated to mean heart rate, which has been known since the first measurements of Schlomka and Reindell [37] in the 1930s, the introduction of this new parameter allows a better differentiation as to the kind of arrhythmia. The immediate practical consequence is demonstrated in the companion paper (Abel et al.) in trained rowers, where this relationship to the respiratory frequency comes into play.

The separation of frequency ranges in the analysis of heart rate variability gives some information on the sympathetic versus vagal contribution to pacemaker innervation [7, 32]. It is known from animal experiments and experiments in man with application of atropine and β-blockers that the vagal tone is expressed by variability in the higher frequency range and the sympathetic tone mainly by that in the lower frequency range. Accordingly, a small activation through a mental test decrease the rhythmicity in the higher but not that in the lower frequency range. Interestingly, the return to control values needs much longer time for the rhythmicity than for the mean values of heart rate (see Abel et al., this volume).

Vasomotricity

Until now we have used as an indicator the finger plethysmogram, recorded with a simple photoelectric device. Vasoconstriction is manifested here in two ways: (a) by the overall variations in mean value, which are quite wide in normal subjects, and (b) by the amplitude of the volume pulse. We record these separately using an additional trace with the high-pass filtered signal. The nonfiltered curve can be analysed according to its rhythmic pattern, where striking differences can be seen, for example, in some groups of renal patients. In the analysis of nonfiltered curves we must be careful since we have learned from animal experiments that peripheral vasomotricity does not mirror exactly the neuronal vasomotor innervation but the interplay of the latter with the rhythmicity of the vascular smooth muscle [38]. In addition, this is dependent on arterial blood pressure. Methods for discrimination among these three factors must be elaborated. Comparison with intra-arterial blood pressure in surgical patients will be helpful in the interpretation of the interpretation of the plethysmogram.

The high-pass filtered and expanded recording of the volume pulse represents an extremely sensitive measure for every small vasoconstrictor, i.e. for sympathetic activation. With low mental load this parameter is the first to change among all the cardiorespiratory parameters that we measure. Thus, this output from the nervous system shows up to now the lowest threshold for activation. Its higher sensitivity compared with the respiratory reactions confirms the general rule of the anticipation of general activation by vegetative innervation.

Respiratory Trace

Following the principle of remaining completely noninvasive, we avoid mouthpieces, nose clips, masks, etc. and record the changes in the circumferences of thorax and abdomen. Surprisingly, the on-line computed sum of both signals shows a good linear relation to tidal volume in the calibration with a spirometer after the individual experiment. The respiratory trace tells us much more than the usual measures of respiratory volume and frequency when we perform an automatic respiratory pattern analysis. This analysis allows the separation of different partial components of the respiratory motor act which mirror the partial components of central respiratory rhythmogenesis and drive. In this analysis the mean inspiratory air flow is a good indicator for the respiratory drive, whereas the period durations of inspiration and expiration express the intracentral timing mechanisms. In this way we could show that tidal volume is not a primary parameter but the resultant of inspiratory velocity and inspiratory time, quite in agreement with the modern theories of respiratory rhythmogenesis based on animal experiments. Thus the respiratory pattern analysis allows discrimination between the effects of behavioural drives and those of chemical drives in humans [34]. Abel et al. show a practical example of such an analysis. Together with the noninvasive registration of end-tidal P_{CO_2}, we can obtain insight into the actual relative importance of chemical feedback regulation and so-called behavioural influences. Closer inspection of the respiratory trace also gave initial indications of vegatative patterns in the form of single deep breaths, which in their time courses are characteristic of the individual subject where the special type is maintained over weeks. This might be a first step in the direction of recognition of individuality.

Cardiorespiratory Interrelations in Man

Without going into detail, we can already state that there are many indications that cardiorespiratory control also in man is exerted by a common network, where similar rhythms or reaction patterns can be found in partial parameters of both systems, whereas other parameters of the same system behave differently. This is in agreement with the model of a common network, as shown previously, which is not rigidly organized in the form of separate specific centers.

Prospects for Future Developments

This review of revised physiological concepts of the cardiorespiratory control system and of the practical possibilities for applying them to man, together with the first results of such an approach, has shown that nature provides us with many easily obtainable signals about the state of these control systems, and moreover, about the balance in vegetative innervations. The task that we now confront is to apply the concepts, to record the accessible variables, and to learn to read the message of the body – and partly also of the mind – which is hidden in the signals. This principle can be applied in the healthy human being, in altered states of physical training or environmental conditions and in patients during diseased states, surgery, anaesthesia, recovery and rehabilitation. In view of the modern state of recording techniques and computer evaluation, it would be inexcusable not to use the tools avaible for innovative improvements of physiological and clinical methodology.

References

1. Abel HH, Klüßendorf D, Koepchen HP (1985) Pattern of somato-vegetative innervation in healthy humans during rest and mental load. Pflugers Arch [Suppl] 403:R51
2. Abel HH, Klüßendorf D, Koepchen HP (1985) Interaction of behavioural and chemical drive in ventilation and respiratory pattern during and after mental load in human subjects. Pflugers Arch [Suppl 2] 405:R50
3. Abel HH, Klüßendorf D, Krause R, Koepchen HP (1985) Are enhanced respiratory oscillations of heart rate at rest in endurance-trained sportsmen related primarily to increased vagal tone? J Interdiscip Cycle Res 16:240
4. Abel HH, Klüßendorf D, Koepchen HP (1986) Relations between spontaneous events in autonomic and respiratory innervation patterns in man. Pflugers Arch [Suppl] 406:R22
5. Angelone A, Coulter NA Jr (1964) Respiratory sinus arrhythmia: a frequency dependent phenomenon. J Appl Physiol 19:479–482
6. Cohen MI (1979) Neurogenesis of respiratory rhythm in the mammal. Physiol Rev 59(4):1105–1173
7. Eckoldt K, Braun W, Lange V (1985) Heart rate at different autonomic states. J Interdiscip Cycle Res 16(4):252–253
8. Euler C von (1986) Brain stem mechanisms for generation and control of breathing pattern. In: Fishman AP, Cherniack NS, Widdicombe JG, Geiger SR (eds) The respiratory system. American Physiological Society, Bethesda, pp 1–67 (Handbook of physiology, sect 3, vol 2)
9. Gauer OH, Henry JP (1963) Circulatory basis of fluid volume control. Physiol Rev 43:423–481
10. Golenhofen K, Hildebrandt G (1957) Über spontanrhythmische Schwankungen der Muskeldurchblutung des Menschen. Z Kreislaufforsch 46:257
11. Heinemann H, Stock G, Schaefer H (1973) Temporal correlations of responses in blood pressure and motor reaction under electrical stimulation of limbic structures in unanaesthetized, unrestrained cats. Pflugers Arch 343:27–40
12. Kenner T (1971) Dynamic control of flow and pressure in the circulation. Kybernetik 9:215
13. Koch E (1931) Die reflektorische Selbststeuerung des Kreislaufs. Steinkopff Dresden
14. Koepchen HP (1962) Die Blutdruckrhythmik. Steinkopff, Darmstadt
15. Koepchen HP (1969) Vegetative-somatic relationships in single neuron activity in the lower brain stem. In: Evans CR, Mulholland TB (eds) Attention in neurophysiology. Butterworth, London, pp 83–99
16. Koepchen HP (1972) Kreislaufregulation. In: Gauer O, Kramer K, Jung R (eds) Physiologie des Menschen, vol 3. Urban and Schwarzenberg, Munich, pp 327–406
17. Koepchen HP (1975) Atmungsregulation. In: Gauer O, Kramer K, Jung R (eds) Physiologie des Menschen, 2nd edn, vol 6. Urban and Schwarzenberg, Munich, pp 163–310
18. Koepchen HP (1983) Respiratory and cardiovascular "centres": functional entirety or separate structures. In: Schlaefke ME, Koepchen HP, See WR (eds) Central neurone environment and the control systems of breathing and circulation. Springer, Berlin Heidelberg New York, pp 221–237
19. Koepchen HP (1984) History of studies and concepts of blood pressure waves. In: Miyakawa K, Koepchen HP, Polosa C (eds) Mechanisms of blood pressure waves. Japan Scientific Societies, Tokyo; Springer, Berlin Heidelberg New York, pp 3–23
20. Koepchen HP, Lazar H, Borchert J (1974) On the role of the nucleus infrasolitarius in the determination of respiratory periodicity. Proc IUPS Vol. IX, p. 81
21. Koepchen HP, Langhorst P, Seller H (1975) The problem of identification of autonomic neurons in the lower brain stem. Brain Res 87:373–393
22. Koepchen HP, Hilton SM, Trzebski A (eds) (1980) Central interaction between respiratory and cardiovascular control systems. Springer, Berlin Heidelberg New York
23. Koepchen HP, Klüßendorf D, Sommer D (1981) Neurophysiological background of central neural cardiovascular-respiratory coordination: basic remarks and experimental approach. J Auton Nerv Syst 3:335–368
24. Koepchen HP, Abel HH, Klüßendorf D (1985) Heart-rate dynamics in healthy humans before, during and after a mental test. Pflugers Arch [Suppl 2] 405:R50

25. Koepchen HP, Lazar H, Klüßendorf D, Hukuhara T (1986) Medullary apneusis by lesions and cooling in the ventrolateral solitary tract region and genesis of respiratory rhythm. J Auton Nerv Syst, (Suppl. pp. 63–69)
26. Langhorst P, Schulz B, Schulz G, Lambertz M (1983) Reticular formation of the lower brainstem. A common system for cardiorespiratory and somatomotor functions: discharge patterns of neighbouring neurons influenced by cardiovascular and respiratory afferents. J Auton Nerv Syst 9:411–432
27. Litter WA, Honour J, Pugsley DJ, Sleight P (1975) Continuous recording of direct arterial pressure in unrestricted patients. Its role in the diagnosis and management of high blood pressure. Circulation 51:1101–1106
28. Loeschcke HH (1960) Homoiostase des arteriellen CO_2-Druckes und Anpassung der Lungenventilation an den Stoffwechsel als Leistungen eines Regelsystems. Klin Wochenschr 38:366
29. Loeschcke HH (1982) Central chemosensitivity and reaction theory. J Physiol 332:1–24
30. Loeschcke HH, Koepchen HP (1958) Versuch zur Lokalisation des Angriffsortes der Atmungs- und Kreislaufwirkung von Novocain im Liquour cerebrospinalis. Pflugers Arch 266:628–641
31. Magoun HW (1950) Caudal and cephalic influences of the brain stem reticular formation. Physiol Rev 30:459–474
32. Malliani A, Lombardi F, Pagani M, Cerutti S (1986) The problem of approaching the sympathetic and vagal "tone". J Auton Nerv Syst, Suppl. pp. 191–196
33. Mancia G, Ferrari A, Gregorini L, Parati G, Pomidossi G, Bertinieri G, Grassi G et al. (1983) Blood pressure and heart rate variabilities in normotensive and hypertensive human beings. Circ Res 33:96–104
34. Mittelstaedt H (ed) (1954) Regelungsvorgänge in der Biologie. Oldenbourg, Darmstadt
35. Richter DW (1982) Generation and maintenance of respiratory rhythm. J Exp Biol 100:93–107
36. Schlaefke ME (1981) Central chemosensitivity: a respiratory drive. Rev Physiol Biochem Pharmacol 90:171–244
37. Schlomka G, Reindell H (1936) Untersuchungen über die physiologische Unregelmäßigkeit des Herzschlages. Z Kreislaufforsch 28:473–492
38. Seller H, Langhorst P, Polster J, Koepchen HP (1967) Zeitliche Eigenschaften der Vasomotorik. II. Erscheinungsformen und Entstehung spontaner und nervös induzierter Gefäßrhythmen. Pflugers Arch 296:110–132
39. Wagner R (1954) Beispiele und Probleme biologischer Regelung. Thieme, Stuttgart
40. Weigelt H, Lübbers DW (1984) The fine adjustment of capillary blood flow through excitation of the capillary wall. In: Lübbers DW, Acker H, Leniger-Follert E, Goldstick TK (eds) Oxygen transport to tissue V. Plenum, New York, pp 731–737

New Approach to Analysing the Neurovegetative State in Man*

H.-H. Abel, D. Klüssendorf and H.P. Koepchen

Introduction

The aim of the present work is to seek evidence for the applicability to humans of concepts concerning the function and coordination of central autonomic control systems that have been derived from animal experiments. We focus our interest on the behaviour of the respiratory control system and the cardiovascular control system, and especially on their mutual interactions both at a state of low activity during rest and at a state of elevated activity under cognitive-motivational influences. The integrative function of central control mechanisms is expressed by parameters which to some extent can also be recorded in man. Using these measurements, we have developed methods of analysis which allow conclusions concerning the organization of central control systems and their mutual interaction.

Methods

Young healthy volunteers participated in two series of experiments, each consisting of three sessions at 1-week intervals. In series A, 19 men (21.2 years ± 0.7) and 10 women (23.2 years ± 0.5) were examined at rest three times for 45 min. In series B, 12 men (22.3 years ± 0.8) and 5 women (24.0 years ± 2.3) were examined according to the following schedule: session 1, recording at rest for 45 min; sessions 2 and 3, recordings for 15 min at rest followed by a 15-min discriminative task (Raven test) and 15-min recovery period. The subjects sat in a comfortable chair. During recordings at rest, they were instructed to relax as much as possible without falling asleep and exposed to low-level background music. During the discriminative task, the volunteers signalled solutions by button pressing, which required only minimal finger movements. Correct responses were rewarded at the end of the experimental run.

Somatomotor and autonomic parameters were measured continuously with non-invasive techniques (Fig. 1). Respiratory movements were monitored by means of a thoracic and an abdominal belt provided with strain gauges. Cali-

*Supported by Hospal Medizin-Technik, Nürnberg, FRG

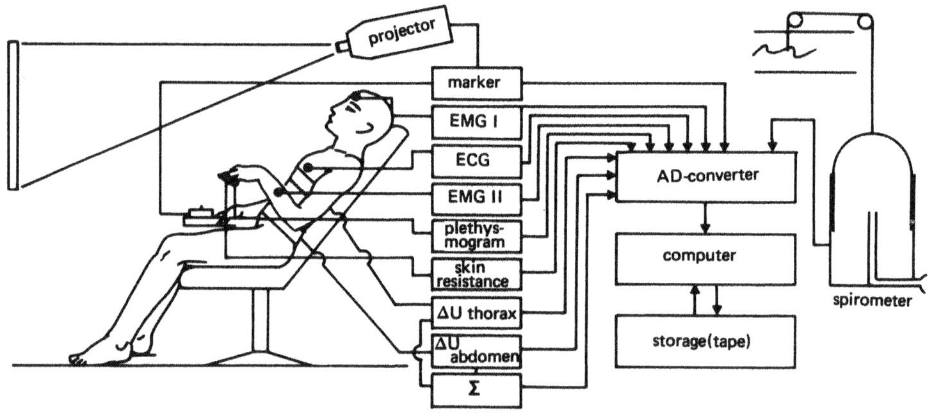

Fig. 1. Experimental equipment

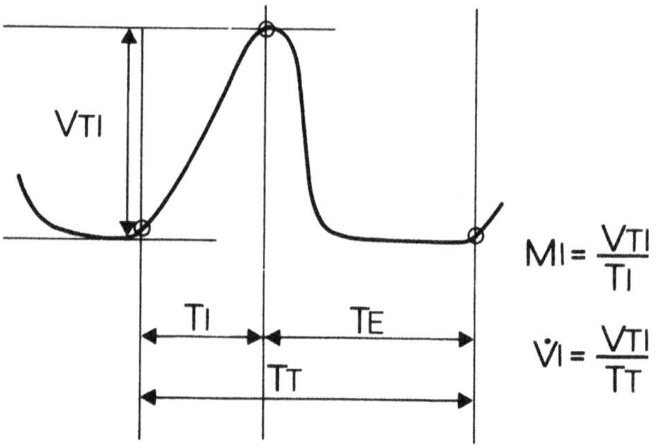

Fig. 2. Scheme of definition of respiratory components

bration was done immediately after the end of the experimental runs by a spirometer. End-tidal P_{CO_2} was recorded simultaneously by means of a tube that ended close to the nostrils. Using the signal obtained by electronic addition of thoracic and abdominal circumferences, respiratory components were computed automatically from breath to breath (Fig. 2). The following parameters and their time courses, mean values and standard deviations were analysed: inspiratory time (T_I), expiratory time (T_E), total respiratory cycle time (T_T), inspired tidal volume (V_{TI}), mean inspiratory flow ($M_I = V_{TI}/T_I$) and inspiratory ventilation ($\dot{V}_I$).

ECG was transformed into an R-wave-synchronous cardiac interbeat interval signal. Mean R-R interval ($\overline{\text{R-R}}$), R-R standard deviation and the power spectrum of R-R intervals were evaluated from consecutive 1-min time sections. For differentiated analysis of heart rate variability, the power in two selected frequency bands was calculated as area under the respective band of the spectrum: area of high-frequency band (HFB, 0.164–0.385 Hz, period duration 6.1 to 2.6 s) and area of mid-frequency band (MFB, 0.049–0.147 Hz, period duration 20.5 to 6.8 s).

Finger pulse amplitude as an indicator of peripheral vasomotor activity derived from the photoelectrically measured finger plethysmogram, finger skin resistance as a measure of peripheral sudomotor activity and EMGs from frontal muscle and forearm muscles were recorded and analysed statistically. All parameters were digitized and stored on tape for further analysis.

Results and Discussion

Relationship Between Heart Rate Variability and Respiratory Frequency

Autonomic innervation of the heart is manifested not only by mean heart rate but also by heart rate variability. It is known that lower mean heart rate is accompanied by higher heart rate variability (Schlomka 1937). In animal experiments and in man a strict quantitative relationship between both parameters was found (Koepchen and Thurau 1959; Katona and Jih 1975; Eckoldt et al. 1980; Fouad et al. 1984). We have examined this relationship in untrained subjects and endurance-trained sportsmen, the latter exhibiting a higher heart rate variability (Schlomka and Reindell 1936). It is well accepted that the magnitude of respiratory sinus arrhythmia is dependent on vagal tone. Since mean heart rate is another indicator of vagal tone, the question arises whether the higher heart rate variability of sportsmen is due to their higher vagal tone at rest maintained at lower heart rates. Evaluations were done from consecutive 1-min time sections of the experimental run.

The positive relationship between mean R-R interval and R-R standard deviation (variability) was confirmed in both groups (Fig. 3). Sportsmen, however, differ from untrained subjects in that they exhibit higher heart rate variability in relation to mean R-R interval, especially in the upper range. At very long mean R-R intervals variability diminishes again. The enhanced variability can be explained mainly by an increase in the magnitude of heart rate fluctuations in the respiratory frequency range as documented by power spectrum analysis (Abel et al. 1985c, 1987c).

Respiratory sinus arrhythmia is known to be dependent on respiratory frequency (Angelone and Coulter 1964; Hirsch and Bishop 1981). Therefore the question arises whether the actual respiratory frequency could influence the relation between heart rate variability and mean R-R interval. This was analysed in untrained and endurance-trained subjects.

Mean total respiratory cycle times were determined in 1-min time sections. The mean values were divided into two groups with high values from 4.5 to 6.2 s and low values from 2.9 to 3.3 s for untrained subjects and sportsmen, respectively. The results are shown as areas in the R-R standard deviation vs. mean R-R interval diagram (Fig. 4). This diagram shows that there is a clear influence of respiratory cycle time on heart rate variability. At longer

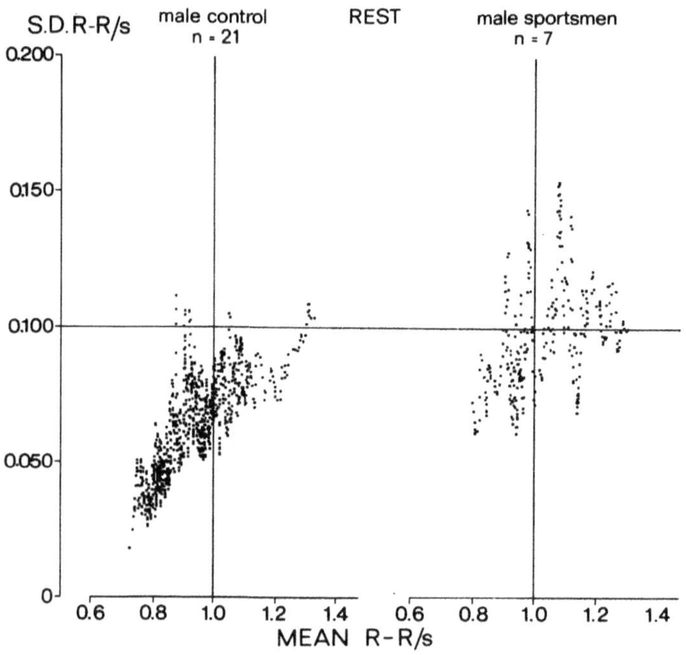

Fig. 3. Relation between R-R standard deviation (*S.D. R-R*) and mean R-R interval (*mean R-R*) from controls (*left*) and sportsmen (*right*); S.D. R-R data smoothed by Hanning window. In sportsmen this relation is shifted towards higher S.D. R-R

respiratory cycle times, heart rate variability in relation to mean R-R interval is distinctly higher in controls and sportsmen. At shorter respiratory cycle times untrained subjects and sportsmen show an identical relation between mean R-R interval and heart rate variability. In contrast, at longer respiratory cycle times heart rate variability in sportsmen is even higher than in untrained subjects. This relation does not continue into the range of the longest mean R-R intervals where variability decreases again, even at longer respiratory cycle times.

These results show that under resting conditions, heart rate variability is related not only to mean heart rate but also to respiratory cycle time. This effect is strongly enhanced in sportsmen. It could be explained as a resonance-like phenomenon when respiratory frequency approaches the lower frequency range of cardiovascular-respiratory rhythmicity. This analysis clearly demonstrates that a judgement about heart rate variability must take into account not only mean heart rate but also respiratory frequency.

Relationship Between Mean Value and Time Structure of Heart Rate

To obtain more precise information about the time structure of heart rate, power spectrum analysis of the heart beat interval was performed, likewise in 1-min time sections. Figure 5 shows a typical power spectrum. The most

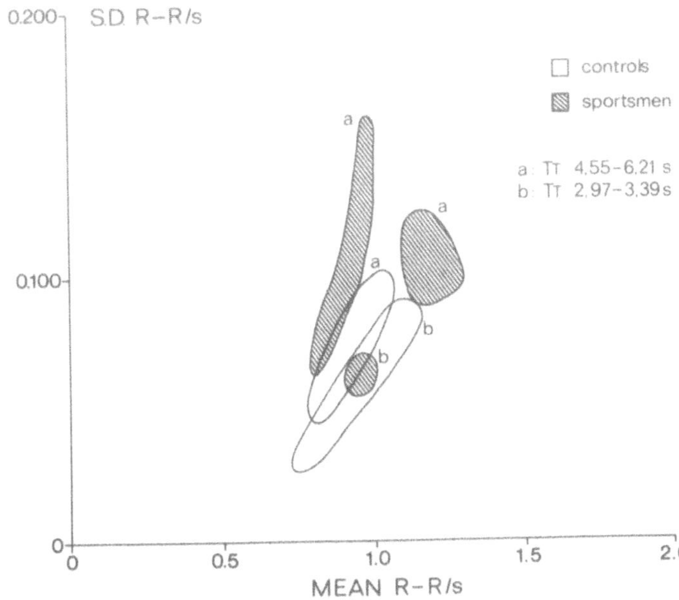

Fig. 4. Relation between R-R standard deviation *(S.D. R-R)* and mean R-R interval *(mean R-R)* from controls and sportsmen during rest, separately analysed for longer (group a: T_T 4.55–6.21 s) and shorter (group b: T_T 2.97–3.39 s) total respiratory cycle times *(T_T)*; S.D. R-R data smoothed by Hanning window. In general, increase in S.D. R-R at longer T_T values. Distinct difference in S.D. R-R between controls and sportsmen are found at the longer but not at the shorter T_T values

prominent peak at about 0.25 Hz represents respiratory modulation of heart rate or respiratory sinus arrhythmia. Peaks at lower frequencies are situated in a frequency range in which circulatory and reticular modulations are also found (Langhorst et al. 1984; Pagani et al. 1986). As a simple measure of heart rate rhythmicity in the different frequency ranges the area under the corresponding section of the power spectrum can be used (Sayers 1973; Akselrod et al. 1981). This kind in evaluation is indicated by the shaded areas in Fig. 5. The power in the HFB is related to the magnitude of heart rate modulation in the preferential respiratory frequency range and can be taken primarily as an indicator of fluctuation of vagal tone. This area totally disappears when atropine is administered. The power in the MFB is related mainly to sympathetic tone (Eckoldt et al. 1985; Pagani et al. 1986).

Figure 6 shows the time courses of mean R-R interval, area of HFB, area of MFB and the quotient of MFB area divided by HFB area in one subject during rest, during the discriminative task and in the following recovery period. A feature in healthy young subjects is the spontaneous increase and decrease of rhythmicity in both frequency ranges, which by itself can be of periodic nature. Areas of HFB and MFB behave partly independently of each other and show only a weak correlation to the mean R-R interval. In this example during the Raven test, mean R-R interval decreases, area of MFB increases and area of HFB decreases. In the recovery period, mean R-R interval has reached the

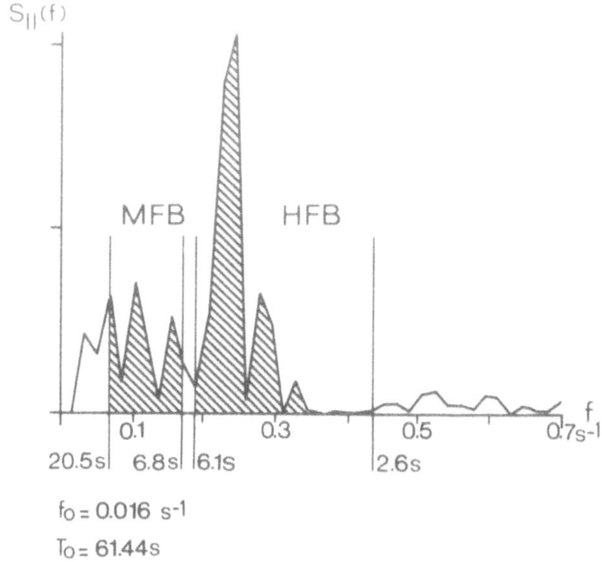

Fig. 5. Power spectrum of cardiac interbeat intervals. Area of the mid-frequency band *(MFB)* and area of the high-frequency band *(HFB)* with respective limits are marked by *hatched areas*

control value, whereas rhythmicity in the HFB is still altered. Thus, rhythmicity in the HFB seems to be a more sensitive indicator of alterations in cardiac innervation than changes in mean value (Abel et al. 1985a; Koepchen et al. 1985a). This is supported by the finding that, in contrast to mean heart rate, rhythmicity changes consistently during cognitive-motivational activation.

Respiratory Pattern Analysis

The discriminative task is an example of the influence of a primarily behavioural drive on ventilation. During the Raven test ventilation increases initially, which leads to a simultaneous decrease in end-tidal P_{CO_2}, in this case caused by increase in tidal volume and respiratory frequency (Fig. 7). Note that increase in tidal volume is largely due to a diminution in functional residual capacity. In the further course ventilation declines and may reach the pre-test value before the end of the task. Due to the simultaneous increase in ventilation and decrease in P_{CO_2} under these conditions the behavioural drive and not P_{CO_2} governs ventilation because P_{CO_2} changes secondarily in the opposite direction (Bülow 1963; Hlastala et al. 1973). Therefore, respiratory changes during the task represent a combination of changes in the behavioural and chemical ventilatory drives.

We have tried to differentiate between the effects of these drives by respiratory pattern analysis. This seemed to be promising because respiratory rhythmogenesis results from interaction of several central mechanisms which have in part been identified and isolated in animal experiments (Richter 1982; Euler 1983; Koepchen et al. 1985b). Since these mechanisms are responsible for components of the respiratory motor act, some indirect conclusions about the

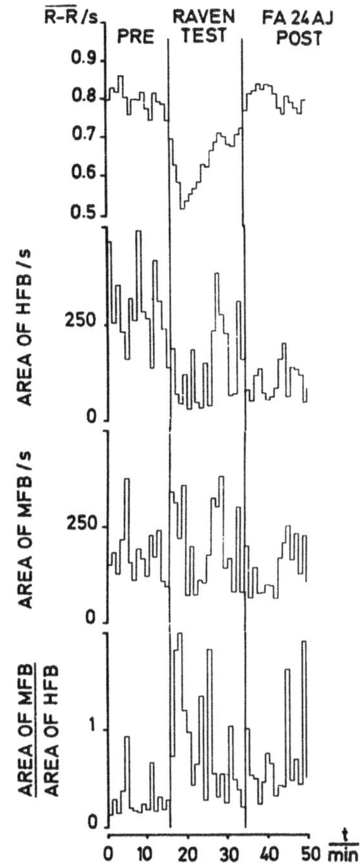

Fig. 6. Time courses of mean R-R interval *(R-R)*, area of high-frequency band *(HFB)*, area of mid-frequency band *(MFB)* and quotient of MFB area divided by HFB area, evaluated in consecutive 1-min time sections during control *(PRE)*, discrimitative task *(RAVEN TEST)* and recovery period *(POST)*. During the task, there is an increase in heart rate and rhythmicity in the MFB, whereas heart rate rhythmicity in the range of respiratory frequency (HFB) is diminished. During the recovery period, heart rate has reached the pre-test control level, while heart rate rhythmicity in the HFB is still below control

activity of single central mechanisms can be drawn also in humans (Abel et al. 1985b, 1987a; Koepchen et al. 1986).

Respiratory frequency increases during the task in all experiments, whereas tidal volume may change in either direction. Increase in mean inspiratory flow and decrease in inspiratory time, however, are consistent in all experiments. This indicates that M_I and T_I are the primary variables on which V_{TI} is dependent. Consequently, we have plotted the primary variables M_I and T_I against each other, whereby lines of constant tidal volume become hyperbolas (Fig. 8). In some other subjects, however, tidal volume decreases, because of a different quantitative relation between increase in mean inspiratory flow and decrease in inspiratory time. The M_I vs. T_I diagram illustrates the possibility of this kind of reaction when the value at the beginning of the task is below the pre-test isovolume hyperbola. In this diagram the most frequent reaction of these subjects, who change respiration during the task, is schematically presented. Starting at the pre-test control value (a) during the initial phase of the task (b), mean inspiratory flow is augmented and inspiratory time is shortened and tidal volume increases. In the further course of the task(c), mean inspiratory flow declines without concomitant changes in inspiratory time, leading to

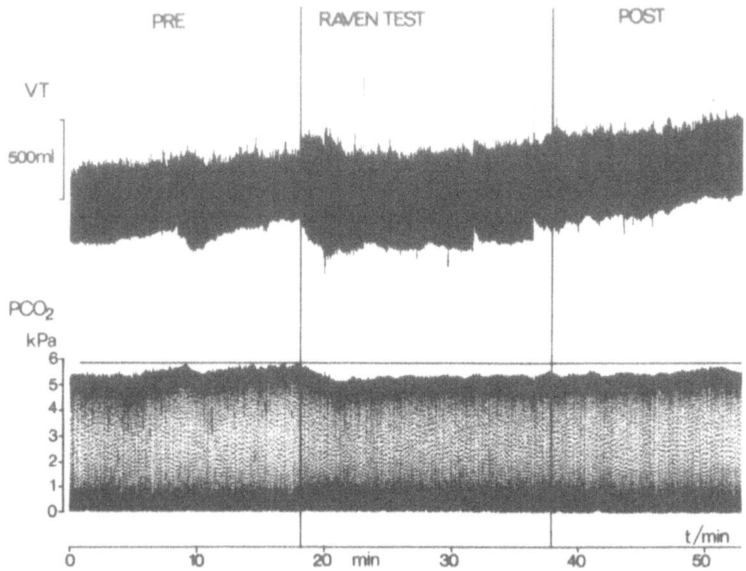

Fig. 7. On the *upper trace* time course of tidal volume, on the *lower trace* time course of end-tidal P_{CO_2} during control *(PRE)*, discriminative task *(RAVEN TEST)* and recovery period *(POST)*. At the beginning of the task, ventilation increases resulting in a secondary change in P_{CO_2} in the opposite direction

reduction in tidal volume. During the early recovery period (d), mean inspiratory flow is below control, whereas inspiratory time has not yet reached the pre-task value. Tidal volume is still reduced. The time course (a, b, c, d in the diagram) describes a hysteresis.

This demonstrates that the factors influencing respiration during the task act in a different manner on the different components of respiratory rhythmogenesis. With respect to the results of animal experiments the role of the two interfering drives in this reaction pattern can be interpreted in the following manner. Mean inspiratory flow is an expression of the activity of the "ramp generator". Inspiratory time depends on the inspiration-terminating mechanism or "switch-off" (Euler 1986). During the initial phase of the task, both the ramp generator and the inspiration-terminating mechanism are activated by the behavioural drive, as shown by an increase in mean inspiratory flow and an decrease in inspiratory time. During the following course of the task, diminution of the chemical drive caused by hyperventilation weakens the activity of the ramp generator only, which results in a decrease in mean inspiratory flow. In contrast, the behavioural drive continues and maintains activation of the inspiration-terminating mechanism, so that inspiratory time remains shortened. During the early recovery period sudden diminution of the behavioural drive acts on the ramp generator in the same direction as the reduction of the chemical drive. Consequently, mean inspiratory flow becomes lower than the control value. The fact that during this period inspiratory time has not yet reached the control value can be explained as a residual effect of a partly

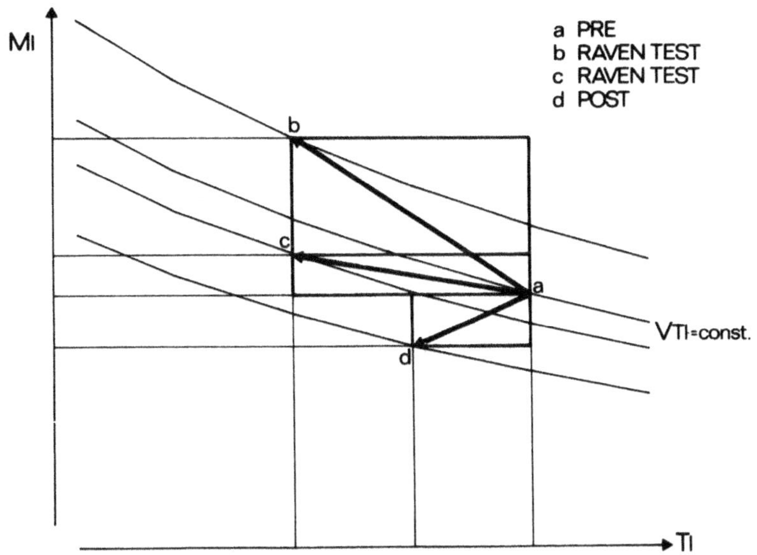

Fig. 8. The primary variables, mean inspiratory flow *(M_I)* and inspiratory time *(T_I)*, plotted against each other. Lines of constant tidal volume become hyperbolas. Reactions of both parameters are represented by vectors. *Vector b,* beginning; *vector c,* last phase of Raven Test; *vector d,* early recovery period. The lengths of the sides of the parallelograms indicate the corresponding changes in the activity of the ramp generator *(vertical axis)* and of the inspiration-terminating mechanism *(horizontal axis)*

maintained behavioural activation. We conclude that in man, too, respiratory rhythmogenesis is the result of interaction of several partial mechanisms which differ in their sensitivity to different respiratory drives. Thus, this new, non-invasive and relatively simple kind of human respiratory pattern analysis can give some indirect information on the interplay of central nervous mechanisms and might be used for the analysis of other physiological and pathophysiological changes in respiration.

Cardiovascular-Respiratory Relations

Direct studies of central nervous activity in animal experiments revealed that cardiovascular and respiratory control systems in the brain stem are tightly coupled and probably in part identical (Koepchen 1983). Therefore, it is interesting to look for indications of central cardiorespiratory coupling under our conditions in man at rest.

Single Deep Breath. The interruption of regular breathing during rest by single deep breaths is a common feature in human respiration. These events are combined with changes in autonomic innervation (Abel et al. 1986). Apart from a consistent increase in heart rate immediately after the spontaneous single deep breath, there are several types of concomitant autonomic changes (Fig. 9).

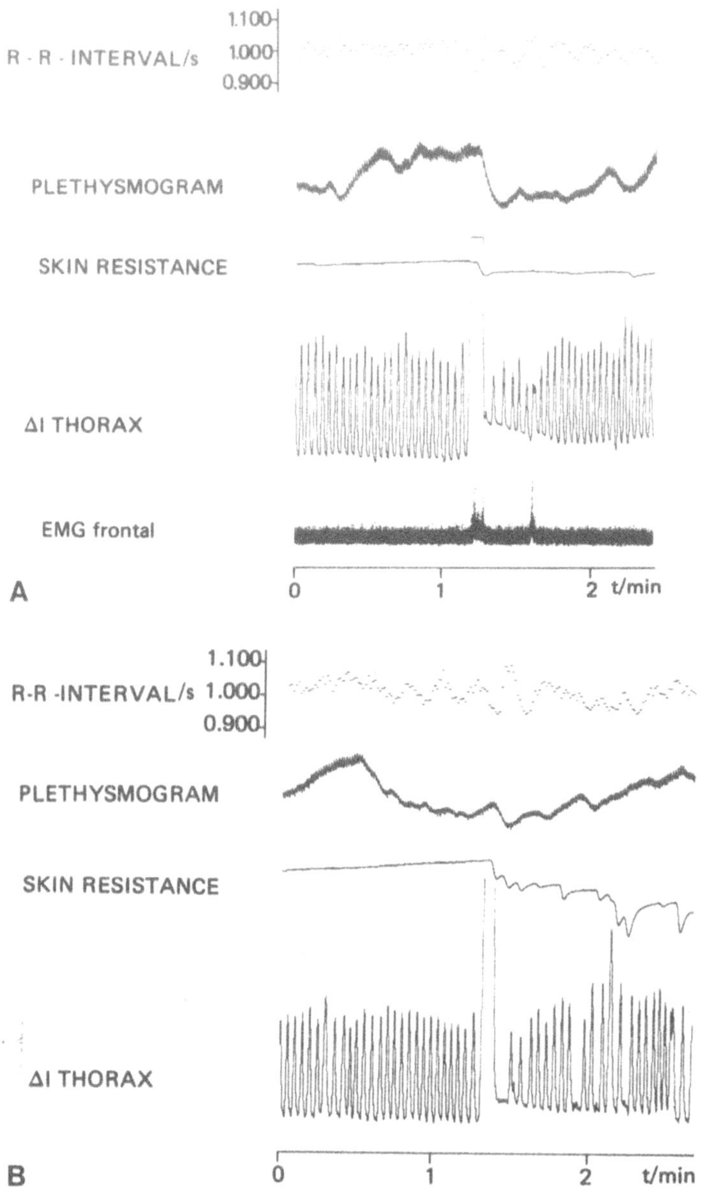

Fig. 9A, B. Two types of behaviour of autonomic and somatomotor parameters at the occurrence of a single deep breath (s.d.b) in the same subject. **A** Changes in all parameters follow secondarily the s.d.b. Note that in heart rate fluctuation a mid-frequency rhythm arises without significant change in mean heart rate. **B** Autonomic innervation precedes a s.d.b. About 1 min before onset of s.d.b. appearance of a mid-frequency rhythm in heart rate and vasoconstriction

First, heart rate frequently becomes rhythmic, with a mean cycle duration of about 10 s starting with the positive phase of heart rate fluctuation without any change in mean value (Fig. 9A). Other autonomic changes following the single deep breath are a decrease in finger pulse amplitude, indicating peripheral vasoconstriction, and a decrease in finger skin resistance, indicating sudomotor activation. Short activation of frontal muscle occurs simultaneously with the single deep breath.

In other cases, changes of cardiovascular innervation can precede a single deep breath (Fig. 9B). About 1 min before its onset, heart rate becomes more rhythmic, and a rhythm appears again in the MFB without change in mean value. At the same time peripheral vasoconstrictor innervation increases, whereas another part of sympathetic efferent innervation, sudomotor innervation, changes just as the single deep breath occurs.

Thus, the single deep breath is part of a general spontaneously occurring activation. These results indicate that in man, too, central control systems are coupled closely but not in a fixed manner, and that the respiratory control system does not necessarily play the leading role.

Parallel Reaction of Cardiovascular and Respiratory Partial Components. The analysis of the time courses of mean values of respiratory and cardiovascular components computed in 1-min time sections during the discriminative task reveals various kinds of coupling between the cardiovascular and respiratory control systems. In subject A (Fig. 10A) a similar time course of cardiovascular parameters, with increase in heart rate and diminution in finger pulse amplitude during the task but only a weak reaction of inspiratory time and mean inspiratory flow are seen. A peculiar kind of stronger coupling between the two control systems is exhibited by subject B (Fig. 10B). Mean R-R interval, inspiratory time, mean inspiratory flow and finger pulse amplitude change at the onset of the task. During the task the time courses of mean R-R interval and inspiratory time, on the one hand, and those of mean inspiratory flow and finger pulse amplitude, on the other, run parallel.

This means that a component of one control system is coupled more strongly with a component of the other control system than with another component of the same system. This supports earlier conclusions drawn from many animal exeriments that the central substrates are not organized according to the principle of special centres for the different peripheral systems.

Individual Characteristics in Spontaneous Patterns

Individual subjects can show constant patterns of neurovegetative parameters over a long time (Myrtek 1984). As a first approach to the question of "neurovegetative individuality", we studied the behaviour of functional residual capacity and tidal volume after a single deep breath and during speech (Abel et al. 1986, 1987b).

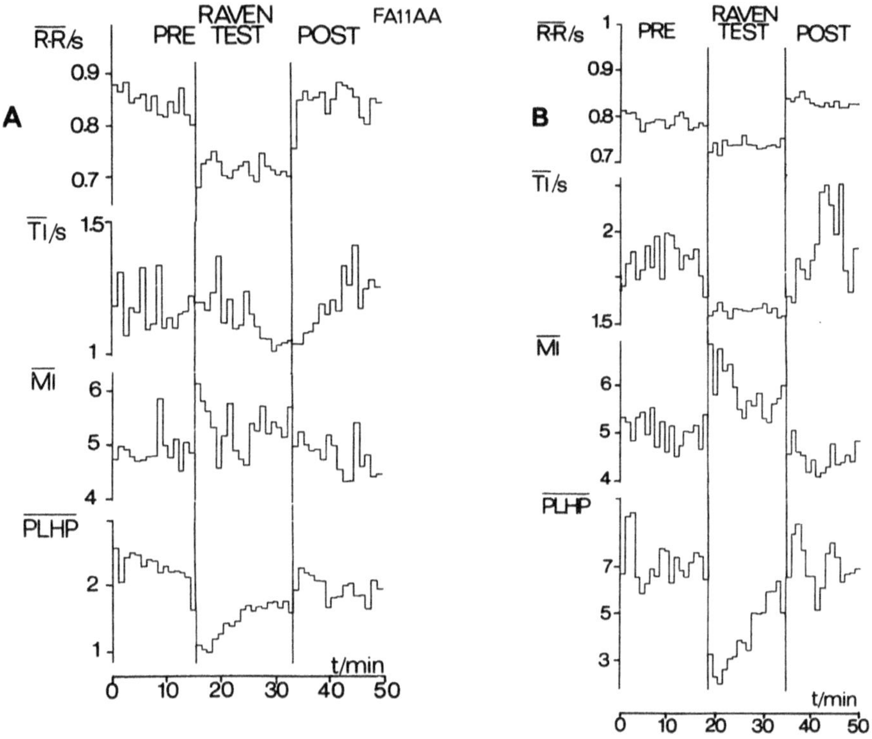

Fig. 10A, B. Time courses of mean R-R interval *($\overline{R\text{-}R}$)*, mean inspiratory time *($\overline{T_I}$)*, mean inspiratory flow *($\overline{M_I}$)* and mean finger pulse amplitude *($\overline{PLHP}$)* of two subjects (**A**, **B**) evaluated from consecutive 1-min time sections during control *(PRE)*, discriminative task *(RAVEN TEST)* and recovery period *(POST)*. Scale of M_I and PLHP in arbitrary units. Subject in **A** shows a similar reaction in R-R and PLHP, but only weak reaction in T_I and M_I during the task. Subject in **B** exhibits a peculiar kind of coupling between cardiovascular and respiratory parameters

Figure 11A shows an example of immediate return of functional residual capacity and tidal volume to control levels after a spontaneously occurring single deep breath. Over a period of 3 weeks respiratory pattern is very similar. In another subject (Fig. 11B) functional residual capacity is elevated after a single deep breath, returning gradually to control. Both types of individual respiratory pattern were stable over the long term of weeks. In another series of experiments investigating respiration during a 15-min speech, one group of subjects showed elevated mean functional residual capacity while another group did not exhibit changes in this respiratory parameter. These reaction types likewise remained constant over weeks.

Further analysis will show whether, apart from the described different time courses of functional residual capacity, characteristic individual patterns can also be found in other neurovegetative parameters and their mutual relations. This may be an access to the characterization of human autonomic individuality.

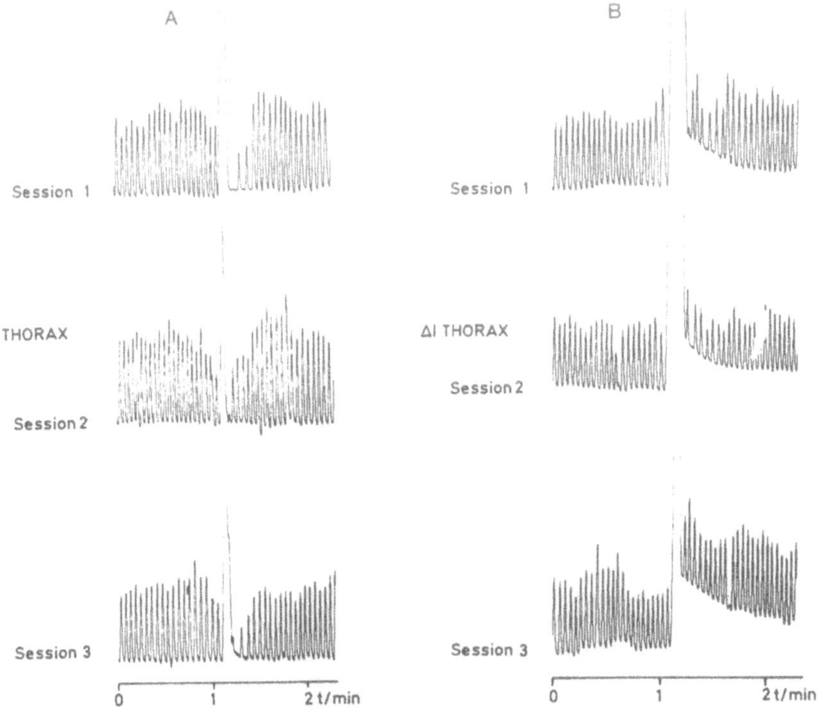

Fig. 11A, B. Long-term stability of two frequently observed individual patterns of functional residual capacity (FRC) following a single deep breath

References

Abel H-H, Klüßendorf D, Koepchen HP (1985a) Pattern of somato-vegetative innervation in healthy humans during rest and mental load. Pflügers Arch 403:R51

Abel H-H, Klüßendorf D, Koepchen HP (1985b) Interaction of behavioural and chemical drive in ventilation and respiratory pattern during and after mental load in human subjects. Pflügers Arch 405:R50

Abel H-H, Klüßendorf D, Krause R, Koepchen HP, Semmler H (1985c) Are enhanced respiratory oscillations of heart rate at rest in endurance-trained sportsmen related primarily to increased vagal tone? J Interdiscip Cycle Res 16:240

Abel H-H, Klüßendorf D, Koepchen HP (1986) Relations between spontaneous events in autonomic and respiratory innervation patterns in man. Pflügers Arch 406:R22

Abel H-H, Mottau B, Klüßendorf D, Koepchen HP (1987a) Differential influence of respiratory drives on central partial mechanisms of respiratory rhythmogenesis in man. Pflügers Arch, 408:S 34

Abel H-H, Mottau B, Klüßendorf D, Koepchen HP (1987b) Pattern of different components of the respiratory cycle and autonomic parameters during speech. In: Sieck G, Gandevia SC, Cameron WE (eds) Respiratory muscles and their neuromotor control. Liss, New York, pp 109–113

Abel H-H, Krause R, Klüßendorf D, Koepchen HP, Semmler H (1987c) Changed relations between cardiorespiratory parameters and heart rate variability in endurance-trained individuals at rest. In: Hildebrandt G, Moog R, Raschke F (eds) Chronobiology & Chronomedicine. Basic Research and Applications. Peter Lang, Frankfurt Bern New York Paris, pp 187–192

Akselrod S, Gordon D, Ubel FA, Shannon DC, Barger AC, Cohen RJ (1981) Power spectrum analysis of heart rate fluctuation: a quantitative probe of beat-to-beat cardiovascular control. Science 213:220–222

Angelone A, Coulter NA Jr (1964) Respiratory sinus arrhythmia: a frequency dependent phenomenon. J Appl Physiol 19:479–482
Bülow K (1963) Respiration and wakefulness in man. Acta Physiol Scand [Suppl] 59:209
Eckoldt K, Pfeifer B, Schubert E (1980) Sympathetic and parasympathetic innervation of the heart at rest and work in man as judged by heart rate and sinus-arrhythmia. In: Koepchen HP, Hilton SM, Trzebski A (eds) Central interaction between respiratory and cardiovascular control systems. Springer, Berlin Heidelberg New York, pp 216–223
Eckoldt K, Braun W, Lange V (1985) Heart rate variability at different autonomic states. J Interdiscip Cycle Res 16:252–253
Euler C von (1983) On the central pattern generator for the basic breathing rhythmicity. J Appl Physiol 55:1647–1659
Euler C von (1986) Brain stem mechanisms for generation and control of breathing pattern. In: Fishman AP, Cherniack NS, Widdicombe JG, Geiger SR (eds) Handbook of physiology, sect. 3: The respiratory system, vol II: Control of breathing, part 1. American Physiological Society, Bethesda, pp 1–67
Fouad FM, Tarazi RC, Ferrario CM, Fighaly S, Alicandri C (1984) Assessment of parasympathetic control of heart rate by a noninvasive method. Am J Physiol 246:H838–H842
Hirsch JA, Bishop B (1981) Respiratory sinus arrhythmia in humans: how breathing pattern modulates heart rate. Am J Physiol 241:H620–H629
Hlastala MP, Wranne B, Lenfant CJ (1983) Cyclical variations in FRC and other respiratory variables in resting man. J Appl Physiol 34:670–676
Katona PG, Jih F (1975) Respiratory sinus arrhythmia: noninvasive measure of parasympathetic cardiac control. J Appl Physiol 39:801–805
Koepchen HP (1983) Respiratory and cardiovascular "centres": functional entirety or separate structures? In: Schläfke ME, Koepchen HP, See WR (eds) Central neurone environment. Springer, Berlin Heidelberg New York, pp 221–237
Koepchen HP, Thurau K (1959) Über die Entstehungsbedingungen der atemsynchronen Schwankungen des Vagustonus (Respiratorische Arrhythmie). Pflügers Arch 269:10–30
Koepchen HP, Abel H-H, Klüßendorf D (1985a) Heart-rate dynamics in healthy humans before, during and after a mental test. Pflügers Arch 405:R50
Koepchen HP, Klüßendorf D, Lazar H, Hukuhara T, Abel H (1985b) Conclusions on respiratory rhythmogenesis drawn from lesion and cooling experiments predominantly in the region of ventrolateral nucleus of solitary tract (vlNTS). In: Bianchi AL, Denavit-Saubié M (eds) Neurogenesis of central respiratory rhythm. MTP Press, Lancaster, pp 77–80
Koepchen HP, Abel H-H, Klüßendorf D, Lazar H, Semmler H (1987) Applicability of concepts of respiratory rhythmogenesis deduced from animal experiments to respiratory control in humans. In: Sieck G, Gandevia SC, Cameron WE (eds) Respiratory muscles and their neuromotor control. Liss, New York, pp 103–107
Langhorst P, Schulz G, Lambertz M (1984) Oscillating neuronal network of the "common brainstem system". In: Miyakawa K, Koepchen HP, Polosa C (eds) Mechanisms of blood pressure waves. Japan Scientific Societies, Tokyo; Springer, Berlin Heidelberg New York, pp 257–275
Myrtek M (1984) Constitutional psychophysiology. Research in review. Academic, Orlando
Pagani M, Lombardi F, Guzzetti S, Rimoldi O, Furlan R, Pizzinelli P, Sandrone G, Malfatto G, Dell'Orto S, Piccaluga E, Turiel M, Baselli G, Cerutti S, Malliani A (1986) Power spectral analysis of heart rate and arterial pressure variabilities as a marker of sympatho-vagal interaction in man and conscious dog. Circ Res 59:178–193
Richter DW (1982) Generation and maintenance of the respiratory rhythm. J Exp Biol 100:93–107
Sayers BMcA (1973) Analysis of heart rate variability. Ergonomics 16:17–32
Schlomka G (1937) Untersuchungen über die physiologische Unregelmäßigkeit des Herzschlages. III. Mitteilung. Über die Abhängigkeit der respiratorischen (Ruhe-)Arrhythmie von der Schlagfrequenz und vom Lebensalter. Z Kreislaufforsch 29:510–524
Schlomka G, Reindell H (1936) Untersuchungen über die physiologische Unregelmäßigkeit des Herzschlages. Z Kreislaufforsch 28:473–492

Rhythms of Eating and Drinking in Sportsmen and Possible Applications in Space Medicine

K. A. Kirsch and L. Röcker

Introduction

Endurance athletes and astronauts share one feature after their exposure to exercise or weightlessness: their body weight (BW) is reduced by 3%–5%, mainly due to a negative salt and water balance. Weightlessness in this connection refers to that of short-term space flights lasting not more than 5–10 days. There is good evidence that under both circumstances BW reduction is due mainly to losses from the extracellular space [10, 12, 13]. Therefore, the control mechanisms such as the osmotic and volume control mechanisms must come into play to correct the deficits [2, 3, 6]. Here the antidiuretic hormone system and the renin-angiotensin system should be mentioned because they, too, are heavily involved.

We have investigated the possible role of the osmotic and volume control systems in the recovery period after exercise [8] and more recently in astronauts [10, 11]. Attention was payed to the time course of the central venous pressure (CVP) after these events; in addition, in the athletes we studied the input side, i.e., the eating and drinking pattern. (For methodological details see [4, 5, 8, 9]).

There were indeed many similarities between both dehydration models, and a comparison is therefore tempting and attractive, but such a comparison should nevertheless be carried out very carefully and cautiously.

Results and Comments

Observing the BW changes of endurance runners over several days (Fig. 1), one can notice a fluctuation of more than 3%. For instance, after a bout of exercise BW is reduced, but soon (mostly before the night rest) BW returns to control level. An exception is the cyclist seen in the third row of the figure; his BW remained at a lowered level for 24 h, and he even started the next exercise thus. However, his BW later returned to the control level. Such a pattern is well known particularly after severe, exhaustive exercise. Under these circumstances deficits are maintained for about 48 h before strong repair mechanisms

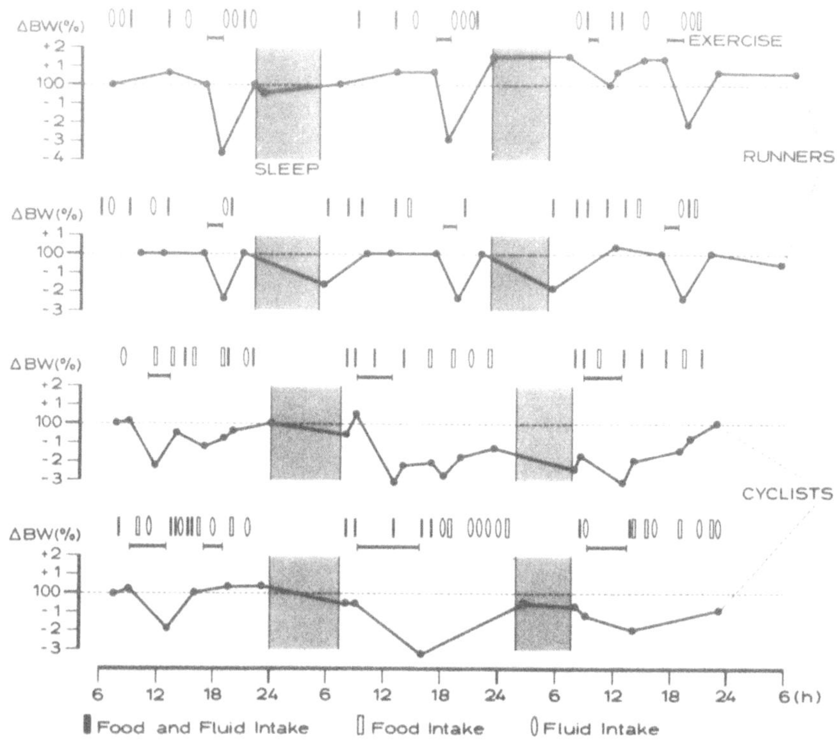

Fig. 1. Changes in body weight *(BW)* in runners and cyclists observed over a period of 3 days. The events of food and fluid intakes are marked above each trace. The bouts of exercise are marked as *horizontal bars*. Notice the frequent food and fluid intakes

come into play correcting the deviation. This example is particularly instructive because in astronauts similar circumstances probably prevail around the landing phase, especially in the 48 h after landing. This phase is one of extreme stress, and many other circumstances prevent an adequate food and fluid intake.

Figure 2 shows the changes in BW early in flight and during recovery for the crew of Skylab 3 [13]. This pattern of initial rapid weight loss and rapid postflight recovery suggests that much of the weight loss associated with space flight is due to loss of fluids. The slower weight loss which persisted later in the long mission will not be discussed here. On the average, postflight weight losses were some 3%–4% of preflight weights.

It is clear that repair of these deficits, regardless of their origin, starts on the input side. Therefore, the eating and drinking pattern of athletes is of particular interest. In general, a high frequency of eating and drinking can be seen in athletes. This is known as a nibbling pattern (Fig. 3). However, these events are not distributed evenly throughout the day. The intervals between eating and drinking are short in the morning and even shorter after exercise in the evening. Longer intervals between eating and drinking are seen regularly in the

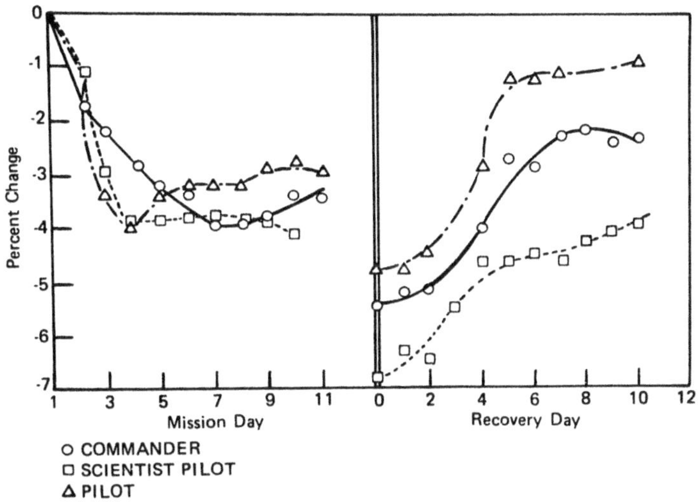

Fig. 2. Changes in body weight early in flight on Skylab 3 and during recovery. [From 13]

early afternoon or before exercise. These periods are called "silent periods" because the drive to sleep is greater than the drives to eat and drink regardless of whether a deficit exists or not. It was also observed that 50% or more of all the fluid intake took place between 6 and 10 P.M. Even in the cases in which they occurred early in the day, as in cyclists, the deficits were carried throughout the day before a restoration took place in the evening. As far as drinking is concerned, a circadian pattern was clear, and in a way described in many animal models [1].

Unfortunately, no such detailed observations are available on astronauts. From our own work with the astronauts during the Spacelab 1 Mission [10] and the D-1 Mission [11], we can say that a similar feeding pattern is very likely. The astronauts feel particularly thirsty after landing. However, "social noise" in the first 48 h after landing overshadows many of these very sensitive mechanisms. On the other hand, studies on astronauts revealed another common feature with endurance athletes. In the recovery period after landing they all showed reduced CVP, which returned toward control levels as soon as BW returned to baseline. The same observations were published by our group regarding long distance runners and cyclists [7].

The example in athletes given in Fig. 4 shows the relationship between BW changes and concomitant CVP changes. A reduction in BW of 4%–5% is usually followed by a fall in CVP 3–4 cmH_2O. Since these changes could be observed long after exercise, thermal stress cannot be made responsible; rather, a fluid loss from the vascular bed and a displacement of blood volume from intra- toward extrathoracic compartments seems to be the most likely explanation [4, 5]. Comparing preflight CVP values in an astronaut with those during the flight one can see a decline within the first 24 h (Fig. 5). This is very similar to the pattern described for BW changes in Fig. 2. In the recovery

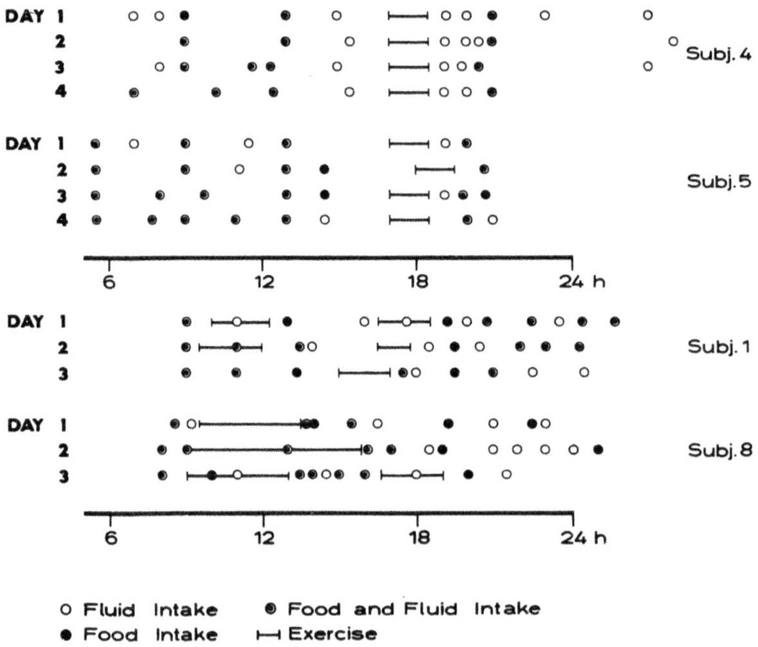

Fig. 3. Daily sequence of eating, drinking and exercise of four endurance exercisers. The runners (subjects 4, 5) were observed on 4 consecutive days, the cyclists (subjects 1, 8) on 3 successive days. Note that the cyclists ate and drank during exercise. Longer intervals between eating and drinking were noticeable in the early afternoon

period CVP is lowered compared to the prelaunch period, however it rises within 3 days after landing. BW was reduced after landing, but the astronaut gained weight in the readaptation phase. This took about 3–4 days.

The two parameters BW and CVP thus seem to be related. This fact should be emphasized because it is well known that intrathoracic receptors located in the right atrium can sense the deficits (Henry-Gauer reflex; [3, 6)] and thereby influence the intake pattern for fluids [2, 8] as well as the output mechanisms via the kidneys [3, 6]. In a deficit situation after exercise a lowered CVP drives the urge to drink as long as the deficit exists. It is true that the concomitant variations in plasma osmolality and the renin-angiotensin system also stimulate the repair mechanisms such as drinking. To our experience, however, CVP level is the most reliable signal for the body [8]. Further experimentation is needed to confirm this conclusion.

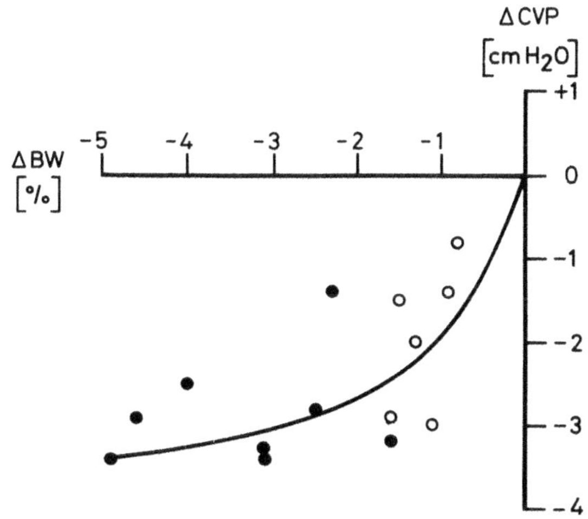

Fig. 4. The changes of central venous pressure *(CVP)* plotted in relation to the changes of body weight *(BW)*

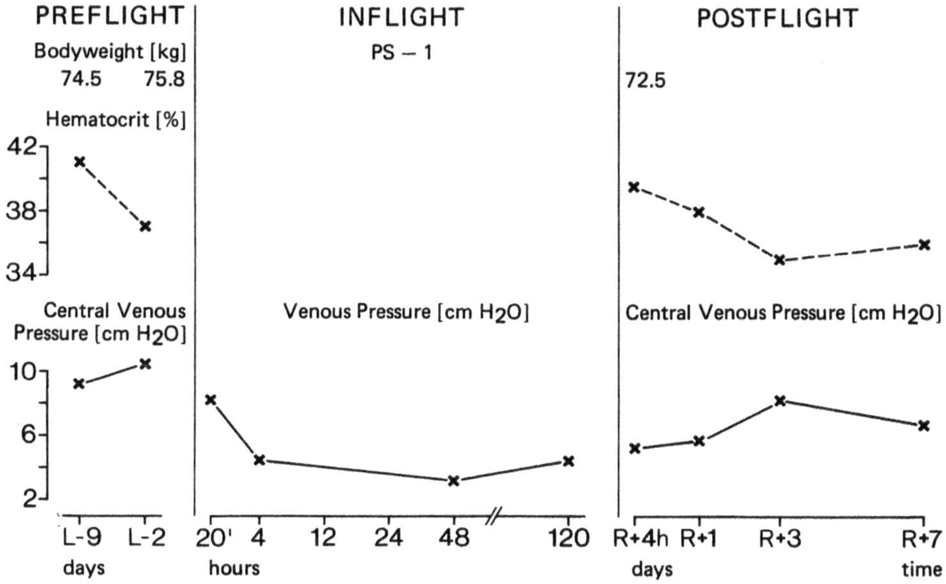

Fig. 5. Time course of hematocrit and venous pressures; pre-, in-, and postflight values from one astronaut (PS-1). Results were obtained from the D-1 Mission. Notice the decline of venous pressure within 4 h after launch. The venous pressure remained lowered until the end of the mission. Within the first 3 days after landing an increase is visible

References

1. Fitzsimons JT (1957) Normal drinking in rats. J Physiol (Paris) 138:39
2. Fitzsimons JT (1975) Thirst and sodium appetite in the regulation of the body fluids. In: Peters G, Fitzsimons JT, Peters-Haeteli L (eds) Control mechanisms of drinking. Springer, Berlin Heidelberg New York, pp 1–7
3. Gauer OH, Henry JP (1963) Circulatory basis of fluid volume control. Physiol Rev 43:423–481
4. Gauer OH, Sieker HO (1956) The continuous recording of central venous pressure changes from an arm vein. Circ Res 4:74–78
5. Gauer OH, Henry JP, Sieker HO (1956) Changes in central venous pressure after moderate hemorrhage and transfusion in man. Circ Res 4:79–84
6. Gauer OH, Henry JP, Behn C (1970) The regulation of extracellular fluid volume. Ann. Rev Physiol 32:547–595
7. Kirsch K, Risch W-D, Mund U, Röcker L, Stoboy H (1975) Low pressure system and blood volume regulating hormones after prolonged exercise. In: Howald H, Poortmans JR (eds) Metabolic adaptation to prolonged physical exercise. Birkhäuser, Basel, pp 315–321
8. Kirsch KA, von Ameln H, Wicke HJ (1981) Fluid control mechanisms after exercise dehydration. Eur J Appl Physiol 47:191–196
9. Kirsch KA, von Ameln H (1981) Feeding patterns of endurance athletes. Eur J Appl Physiol 47:197–208
10. Kirsch KA, Röcker L, Gauer OH, Krause R, Leach C, Wicke HJ, Landry R (1984) Venous pressure in man during weightlessness. Science 225:218–219
11. Kirsch K, Haenel F, Röcker L, Wicke HJ (1986) Venous pressure in microgravity. Naturwissenschaften 73:447–449
12. Kozlowski S, Saltin B (1964) Effect of sweat loss and body fluids. J Appl Physiol 19:1119
13. Thornton WE, Ord J (1977) Physiological mass measurements in Skylab. In: Johnson RF, Dietlein LF (eds) Biomedical results from Skylab. US Government Printing Office, Washington (NASA SP-377)

Closed-Circuit System

The Closed-Circuit – A Perfect Anesthesia and Monitoring Unit

R. Droh

Induction of Anesthesia

Following premedication as with open, half-open, and half-closed systems, we generally induce anesthesia for adults with thiopental: occasionally we also administer Ketanest or Hypnomidate. In every case we use the full prepared dosage (e.g., 500 mg thiopental in a 20-ml syringe). In our opinion no benefit accrues to the patient when, in the induction phase, the anesthesia is continued with another agent (e.g., halothane) instead of using the thiopental, Ketanest, or Hypnomidate remaining in the syringe. In this way we gain time for the induction of the volatile anesthetic, while also reducing costs and problems of disposal.

Until the patient is intubated, we respirate, as usual, manually using a mask. We do this either with pure O_2 (approximately 6 l/min) or with an O_2 and N_2O mixture (1.5:4.5 l/min) with the pressure-relief valve half closed while adding the volatile anesthetic via a vaporizer. Inspiratory and end-expiratory concentrations of O_2, CO_2, N_2O, and volatile anesthetic are monitored continually. With such high gas flow rates the vaporizer control is set to 0.3–0.7 vol.% for halothane. For the vaporization of halothane in closed systems we prefer the vapor (Dräger, Lübeck, FRG), which has an upper dose limit of 7 vol.% (Fig. 1). The same effect may be achieved with conventional vaporizers having an upper limit of 4 vol.% when two units are connected in series. We do not use Low's injection technique for volatile anesthetics; with this method, each injection of the volatile substance into the breathing tube results in a very high vapor concentration – practically a cloud. The concentration falls only slowly, in a stepwise, breath-for-breath manner, with accompanying parallel changes in pulse and blood pressure. With the help of the much more exact dosaging possibilities provided by the Dräger vapor and by continual monitoring of anesthetic and gas concentrations, hyperventilation-induced cardiac and circulatory reactions, as well as over- and under-dosage, may be completely avoided. Further more, future improvements in the apparatus for controlling and monitoring the concentrations and amounts of gases and vapors are to be expected.

In order to fill the system quickly to the desired concentration of anesthetic gases and vapors after intubation, a gas flow of 1.5 l O_2/min, 4.5 l N_2O, and a vaporizer setting of 0.3–0.7 vol.% halothane is usually chosen. The system

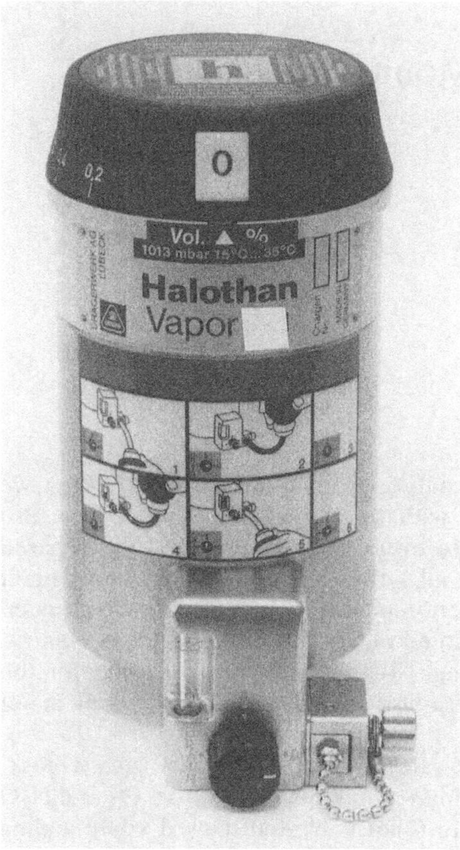

Fig. 1. Halothane Vapor unit (Dräger) for the vaporization of halothane, with an upper limit of 7 vol.%

remains half closed until the desired concentrations, in this case 25 vol.% O_2, almost 75 vol.% N_2O, and 0.3–0.7 vol.% halothane, are reached. With the high gas flows used (1.5 l O_2/min and 4.5 l N_2O/min) this occurs within a few minutes. The system is then closed, with the reservoir bag normally full (0 cmH_2O). It is important that all parts of the system, from the trachea (control for proper sealing of the endotracheal tube) to the connecting pieces, are gas tight (see also Fig. 12). Needless to say, it is imperative that the system be tested in this regard before each use.

As soon as the system is closed, the influx of fresh gases is adjusted to the patients's consumption or uptake. The O_2 requirement of the patient at rest is first estimated as 3.5 ml O_2/kg. This is in terms of an adjusted body weight, which here is defined for men as height in cm minus 100 and for women as height in cm minus 100 minus 10%. For example in a man 180 cm tall and weighting 100 kg we take 180 minus 100 = 80 kg as the adjusted body weight for calculating the basal O_2 requirement. Thus, $80 \times 3.5 = 280$ ml O_2 required per minute. The rotameter is then set at 280 ml O_2/min for fresh gas inflow as soon as the system is filled and closed as described above.

The N_2O requirement is calculated at first as 1.5 ml/kg adjusted body weight. Adjusted body weight is calcualted here as for O_2. As it is physiologically inert, N_2O is not metabolized, but it must be continually replaced at approximately this rate as it diffuses through the skin into the air and through the peritoneum when the abdomen is opened. For the man in the above example, whose adjusted body weight is 80 kg, an N_2O requirement of $80 \times 1.5 = 120$ ml N_2O/min is calculated. In the case of slender patients, if the actual weight is less than the adjusted body weight calculated from height, we take the actual weight and not the adjusted body weight for all calculations of O_2 requirement and N_2O uptake, as well as for volatile anesthetics.

In general, we take body weight minus the weight of body fat for the calculations of fresh gas inflow, as this is the mass which takes part in basal metabolism and is therefore the mass significant for our calculations.

Maintenance of Anesthesia

Addition of Fresh O_2

The mixture of anesthetic with gases is adjusted in the following manner:

1. At present still using a rotameter with one flow per minute (Fig. 2); in the future, certainly using computer-controlled flowmeters with extremely fast-acting valves which can inject gases from breath to breath.
2. Via gas concentration measurements (Fig. 3) for O_2 alone, using commercially available O_2 measuring devices. With these devices the O_2 concentration is measured and the remainder is a mixture of unknown proportions (e.g., of N_2O and N_2).
3. As an alternative, using commercially available measuring devices (Fig. 4) which allow the simultaneous measurement of the concentration of several gases, such as N_2O, O_2, CO_2, and volatile anesthetics. In the future such measurement devices will surely be integrated with computer-controlled flowmeters so that not only the concentrations, but also the amounts of each gas taken up from breath to breath may be determined.

As long as the normal air mixture (79 vol.% N_2 + 21 vol.% O_2) present in the patient's lungs is not washed out with pure O_2, either before or during the induction of anesthesia, the N_2 remaining in the lungs mixes with the mixture of O_2 and N_2O in the closed-circuit system. For example, if the closed-circuit system has a volume of 5 l and the patient also has a total capacity of 5 l, 79 vol.% N_2 of the patient's 5 l, roughly 4000 ml N_2 (and 1000 ml O_2), mixes with the 5 l N_2O + O_2 anesthetic mixture. If the latter mixture contains, for example, 75 vol.% N_2O and 25 vol.% O_2, the 5 l contains 3750 ml N_2O and 1250 ml O_2. This means that the 10 l total gas volume (total capacity of the lungs plus the volume of the closed-circuit system) contains 37.5 vol.% N_2O, 40 vol.% N_2, and 22.5 vol.% O_2 after mixing of the gases.

If anesthesia was induced using a half-open or half-closed system and pure O_2, the free N_2 in the lungs can be washed out, and the N_2 concentration thereafter is correspondingly low once the system is closed. In this case the only

Fig. 2. Rotameters (Dräger) for the dosing of O_2 and N_2O and having ranges between 20 and 10 000 ml/min

N_2 present in the system is that given off from the body tissues: approximately 1 vol.% N_2 per hour. We prefer an inspiratory O_2 concentration of 25 vol.% and an expiratory concentration of 21 vol.%. A higher O_2 concentration at the beginning of anesthesia may be washed out with a high gas flow through an open pressure-relief valve until the desired gas concentration is reached. Alternatively one can close off the inflow of O_2 and let only N_2O into the system until the excess O_2 is taken up by the body. In the case, in order that O_2 consumption does not result in the closed-circuit being breathed empty, the fresh gas influx is given solely as N_2O, but in an amount corresponding to: the calculated amount of N_2O plus the calculated amount of O_2.

Thus before anesthesia is begun we determine (a) the anticipated O_2 consumption/min, (b) the anticipated N_2O uptake/min, and (c) the total amount of gas/min resulting from (a) and (b). The total amount of gas/min varies, due to the decrease in O_2 and N_2O requirement during anesthesia and the increase in requirement during induction and termination of anesthesia as well as during reduction of the depth of anesthesia.

In the expiratory part of the system, metabolized O_2 appears as CO_2 (plus H_2O when fat is metabolized) and is absorbed by the soda lime. As CO_2 is bound by the lime, the total amount of gas in the system and systemic gas pressure are reduced. The closed system, then, loses gas at the same rate that

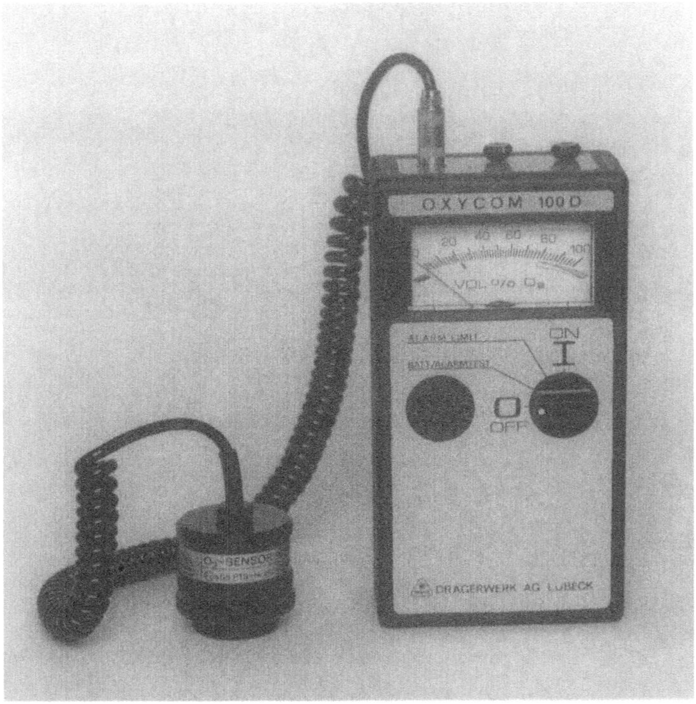

Fig. 3. Oxycom (Dräger) measurement instrument for the determination of O_2 concentration in vol.%.

CO_2 is produced and absorbed, as long as it is not replaced by an equivalent amount of fresh O_2.

If, during anesthesia, the O_2 concentration falls below the desired level, the total fresh gas inflow can be given as pure O_2 until the inspiratory gas mixture again shows the desired O_2 concentration. Alternatively, a high rate of fresh gas inflow (e.g., 6 l/min) can be used with the pressure-relief valve open in order quickly to flush out and refill the system; the system is then closed again (as described above) and the anesthesia continued. It is possible to regulate the O_2 inflow rather finely and continuously over the whole course of the anesthesia by taking the O_2 consumption of the patient to be equal to the expired CO_2. That is, a respiratory quotient (RQ) of 1 is assumed (RQ = CO_2/O_2, for carbohydrate metabolism). In this case, the volume of CO_2 produced per minute is replaced by the same volume of O_2 per minute. This may be achieved (Fig. 5) by use of the Servoventilator 900 C and the Analyser 930 (Siemens).

Since N_2O is not metabolized, its uptake should decrease toward is steady state of zero, depending on the duration of anesthesia. It does not, however, decrease to zero because N_2O diffuses out via the skin, peritoneum, intestinal tract, and those parts of the closed-circuit system made of rubber.

The desired O_2 concentration and O_2 volume in the anesthetic gas mixture can be well maintained, as long as the closed-circuit system is absolutely gas

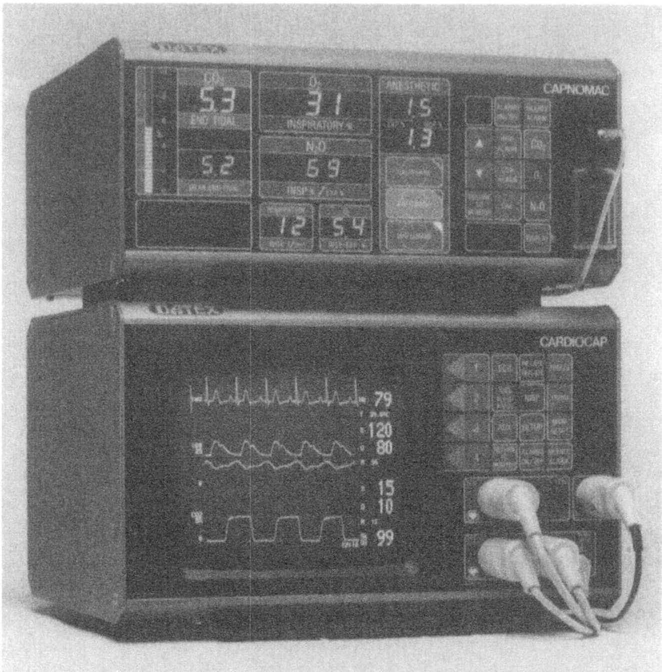

Fig. 4. Normocap (Datex) for measuring concentrations of O_2 and CO_2 in vol.% and Multicap (Datex) for measuring concentrations of O_2, CO_2, N_2O and volatile anesthetics in vol.% combined in Capnomac (Datex)

tight, and the measuring devices function reliably. The time necessary to achieve the desired gas concentrations depends on (a) how high the gas flow of O_2 and N_2O per minute is, (b) how much gas is in the closed-circuit system, and (c) how high the O_2 consumption and N_2O uptake for the particular patient are.

For example, in a young male patient with a body weight of 70 kg and a height of 170 cm, O_2 consumption in the beginning phase of anesthesia would be 70 × 3.5 ml O_2/min = 245 ml O_2/min or 4.08 ml O_2/s; N_2O uptake would be 70 × 1.5 ml N_2O/min ≐ 105 ml N_2O/min or 1.705 ml N_2O/s. The total amount of gases (N_2O + 0_2) would thus be 350 ml/min.

If the O_2 concentration of the breathing mixture is 30 vol.%, but the desired O_2 concentration on the inspiratory side is 24 vol.%, and the total volume of gas circulating in the system and lungs is 10 l, there is an excess of 6 vol.%, or 600 ml, O_2 in the system. This O_2 excess is used up by the patient in 147 s (600 ml O_2 divided by 4.08 ml 0_2/S). This means that in 147 s the O_2 surplus will be consumed such that the initial O_2 concentration of 30 vol.% will have sunk to 24 vol.%, as long as no fresh O_2 is added during this time, and the system is absolutely gas tight.

If this consumption of 600 ml O_2 is not compensated by an inflow of the same amount of N_2O, the total amount of gas in the system is reduced by this 600 ml,

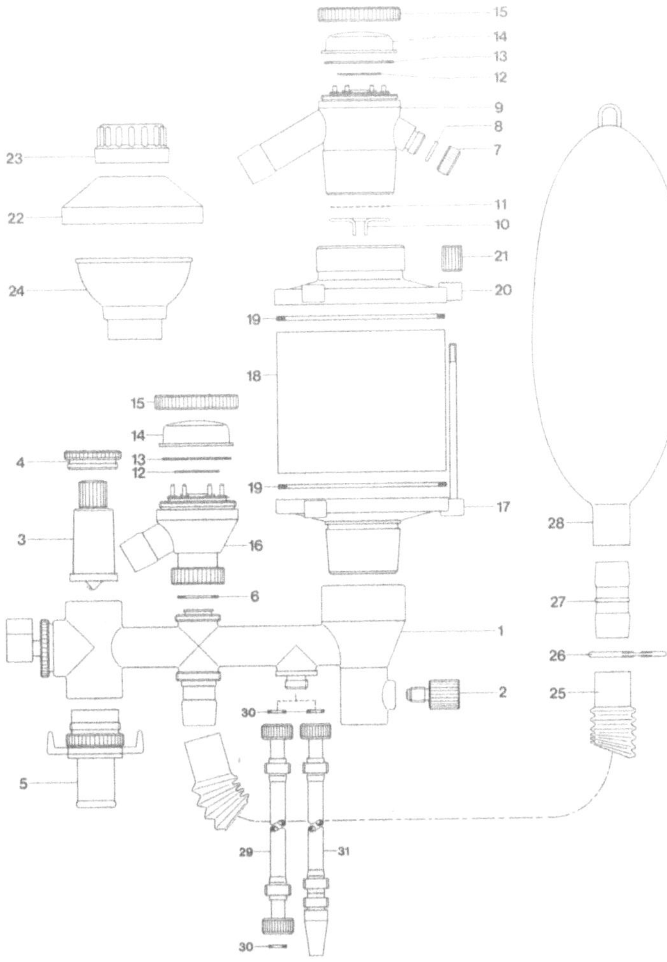

Fig. 5. Individual parts of the Dräger closed-circuit systems 7 and 7a

and the pressure in the system falls accordingly. In this way, 600 ml gas in the system is lost; the effect is greater, the smaller the volume of the system is. Therefore, during this calculated period of 147 s needed for the O_2 concentration to sink from 30 vol.% to 24 vol.% we set an inflow of 350 ml N_2O/min instead of the calculated values of 245 ml O_2/min and 105 ml N_2O/min. After the 147 s we set the inflow of fresh again to the necessary values for O_2 and N_2O.

As this example shows, a sufficiently large volume of total gas in the closed system buffers the system against abrupt changes in gas concentrations and volume. This is also the case for the normal patient at rest when connected to a totally closed and absolutely gas-tight system. On the other hand, there are phases in which the O_2 consumption is increased or decreased.

Increased O_2 Consumption After Fine Regulation of Gas Concentrations and Fresh Gas Inflow

Always assuming that the system is absolutely gas tight, we can determine, once the system is connected to the patient, filled with gases and closed, whether the patient consumes and takes up the amounts of O_2 and N_2O that we have calculated. If with the patient in the above example with inflows of 245 ml O_2/min and 105 ml N_2O/min (total of 350 ml gas inflow/min), the total volume of gas in the system is decreasing, i.e., intrasystemic pressure is falling, this may be due to any one of several causes.

First, the closed-circuit system, including its connection (intubation) with the patient may not be absolutely gas tight; this may result from one or more of the following factors:

- The cuff of the endotracheal tube does not press sufficiently tightly against the tracheal wall.
- The connector between the endotracheal tube and the Y piece either does not fit sufficiently tightly in the tube or does not seal precisely with the Y piece.
- One or more of the various connectors in the system are not absolutely tightly sealed.
- The valves of the system do not close tightly.
- The CO_2 absorption units are not sealed tightly or their connections to the system are not gas tight.
- The breathing tubes are porous, inelastic and do not seal absolutely tightly over the metal connectors of the system.
- The breathing tubes or reservoir bag have become porous so that substantial amounts of gas can leak directly through them.
- The metal parts of the system have developed hair line cracks, or the connectors are leaky, or their sealing surfaces are scratched or otherwise damaged.

Second, there may be an error in the calculations, or the values for body weight and height are wrong.

Third, there may occur any of a number events during induction of anesthesia, such as:

- The patient becomes excited.
- The muscle relaxant (succinyldicholine) elicits strong muscle fibrillations.
- The anesthesia is not deep enough, and the patient has an increased muscle tone with stress-hormone release.
- A tourniquet opens and the oxygen debt for this area must be compensated.
- There is a malfunction of thyroid, hypophysis, or adrenal glands causing increased metabolism (e.g., hyperthyroidism).
- There is developing a malignant hyperthermia or a compartment syndrome.
- The endotracheal tube is improperly positioned (esophagus).

Fourth, gas samples diverted for measuring the amounts and concentrations of CO_2, O_2, N_2O, volatile anesthetics, tidal volume, respiration pressure, gas temperatures, and humidity are not being fully returned to the system.

And, fifth, there may be difficulties, with equipment. The devices for delivering the fresh anesthetic gases (rotameters) may not be set precisely, and either too much or too little of these gases is being added into the system. Another such difficulty is presented by a flowmeter for measuring breath volume which is inaccurate. (Equipment with built-in instruments for computer-assisted measurement of inspiratory and expiratory gas concentrations does not yet allow exact control of gas inflow rates.)

Decreased O_2 Consumption After Fine Regulation of Gas Concentrations and Fresh Gas Inflow

On the other hand, the closed-circuit may become overfilled, that is, the pressure in the system may rise. This can also have several causes.

First, O_2 consumption may have been calculated on an incorrect body weight, height, sex, or age. O_2 requirement diminishes with increasing age, the O_2 requirement of a 70-year-old being about 15% less than that of a 20-year-old.

Second, conditions may develop due to the fact that O_2 requirement at rest is substantially higher than that under anesthesia:

- During artificial respiration the work of breathing does not take place.
- With muscle relaxation (reduction of muscle tone) the O_2 demand for this part of the basal metabolism decreases.
- The body temperature of the patient decreases during general anesthesia. O_2 consumption decreases by 50% at 28°C body temperature.
- O_2 consumption decreases with increasing duration of anesthesia. The longer the anesthetic act pharmacologically on the patient, the more the metabolism is decreased.
- O_2 consumption is reduced with increasing depth of anesthesia. Here too, metabolism is depressed.

Third, O_2 supply to parts of the body may be interrupted due to unperfused areas, vascular blocks, extracorporeal circulation, diminished perfusion or diffusion, as well as shock.

Fourth, leaky measuring instruments in the bypass may pump additional air or O_2 into the system.

Fifth, the delivery apparatus (rotameter or flowmeter) for the anesthetic gases is too inaccurate (measurement error) and may introduce more gas into the system than indicated.

Sixth, metabolic disturbances with diminished metabolism may be present (e.g., hypothyroidism).

A seventh condition is that in which gas exchange in the lungs and/or in body tissues is disturbed (e.g., poisoning).

And, finally, cardiocirculation may be disturbed.

As an example we can contrast the clinical case of a 73-year-old male patient of 80 kg body weight and 171 cm height with the previous calculations for the young male patient. The corrected body weight is calculated as $171 - 100 = 71$ kg. If this patient were young, he would require, at rest, $71 \times 3.5 = 248.5$ ml O_2/min and $71 \times 1.5 = 106.5$ ml N_2O/min, or, at the beginning of anesthesia, a total gas flow of 355 ml/min. Forty minutes after the system was filled, and the desired gas concentrations and required inflows were set, the O_2 consumption of this patient was only 170 ml O_2/min. Since this 73-year-old patient had, in addition, a tourniquet on the leg, his resting O_2 consumption dropped from the calculated value of 248 ml O_2/min to 110 ml O_2/min after 50 min of anesthesia and 40 min of blocked consumption. This reduction in O_2 consumption is a result of the following factors:

- The patient's age (73 years). A reduction of about 15% is to be expected; 15% of 248.5 ml O_2 = 37.3 ml O_2. The resting O_2 consumption for this patient should thus be 211.3 ml.
- The constriction of the right leg by the application of a tourniquet. This blocked off approximately 6% of the corrected body weight; 6% of 211.3 ml O_2/ml = 12.7 ml O_2/min. The calculated O_2 consumption would then be 211.3 ml − 12.7 ml = 198.6 ml O_2/min.
- The general reduction in muscle tone (relaxation) of the striated muscle, including the lack of breathing work (artificial respiration) and the reduced body temperature (eyelid temperature, 32.4°C).
- The duration of anesthesia. The metabolism of the thoracic and abdominal organs as well as that of the brain, which together account for 70% of basal metabolism, was reduced.

In this case there was an overall reduction in O_2 consumption of 56%. Of this, approximately 15% was due to the age of the patient. 6% to the constriction of the tourniquet, and 35% to the reduced body temperature combined with the reduction in muscle tone (relaxation) of the striated muscles and the reduction in metabolism of the internal organs.

O_2 Consumption of Organs

The closed-circuit system also provides the possibility of determining the O_2 debt, or consumption, of regions of the body isolated from the circulation by measuring the increase in O_2 consumption following release of a perfusion block. We can illustrate this with an example using the following concepts:

O_2 $flow_I$ = O_2 flow during a perfusion block
O_2 $flow_{II}$ = O_2 flow after release of the perfusion block up to the point at which the accumulated O_2 debt is compensated
O_2 $flow_{III}$ = O_2 flow before setting the perfusion block
t_a = duration of the perfusion block
t_b = duration of the compensation period following release of the perfusion block

We can define the O_2 debt accumulated during the perfusion block as (O_2 $flow_{III}$ − O_2 $flow_I$) × t_a; and the O_2 debt accumulated during the perfusion block as (O_2 $flow_{II}$ × t_b) − (O_2 $flow_{III}$ × t_b). If the pressure in the system and the level of anesthesia are held constant throughout, it is also possible to calculate the percentage of body weight affected by a perfusion block from the results of O_2 consumption, O_2 production, and the duration of the block. That percentage of the body cut off from the O_2 supply during the block for a period can be expressed as: O_2 debt/time.

When the perfusion block is opened, while tidal volume and respiratory frequency are held constant, the end-expiratory CO_2 concentration (and amount of CO_2) rises, because the lactic acid accumulated in the blocked region during anaerobic metabolism is then released and buffered by the bicarbonate. Since the value of $H^+ \cdot HCO_3/CO_2$ is a constant, the pH of the blood can be held constant by releasing CO_2.

We carry out artificial respiration as normoventilation by holding the end-expiratory concentration of CO_2 constant at 4.8 vol.% (the range for normoventilation is between 4 and 5.6 vol.%). Upon releasing a perfusion block, the end-expiratory concentration as well as the volume of CO_2 rises out of this range if the tidal volume, frequency of respiration, and the amount of O_2 are not changed. If we want to hold the end-expiratory CO_2 at 4.8 vol.% after opening a perfusion block, the patient must be hyperventilated in order to compensate the increased release of CO_2. That is, the tidal volume, O_2 flow, and/or respiratory frequency must be increased. The O_2 flow could be increased, for example, from 247 ml/min to 1000 ml/min. In this way the closed-circuit system could not be "breathed up," and the increase in CO_2 could be held at 4.8 vol.%. This increased O_2 inflow of 1000 ml/min must be held, for example, for 40 s in order that the intra systemic pressure and the end-expiratory CO_2 be held at the values previously set for the patient. As soon as O_2 consumption declines to the calculated normal O_2 requirement (without O_2 debt), the intrasystemic pressure must rise for as long as the incresed O_2 inflow is maintained. Alternatively, the patient could be hyperventilated, in order to prevent the intrasystemic pressure from rising. This, however, would only work for a short time and would, of course, drive the end-expiratory CO_2 concentration below 4.8 vol.%.

An O_2 inflow of 1000 ml/min for 40 s corresponds to 666 ml O_2. During such an increased inflow the O_2 concentration of the breathing gas rises if the respiration parameters are held constant. In order to bring this raised O_2 concentration back down to the initial value, respiration may continue without any fresh gas inflow of O_2 as long as the surplus lasts. The O_2 surplus necessary to compensate the O_2 debt represents then the increased consumption in this period minus the O_2 consumption that would have occurred in a comparable period before the perfusion block. With measurements of intrasystemic pressure, end-expiratory CO_2, amount of CO^2, and total O_2 consumption, including the increase in O_2 consumption following release of the perfusion block, one can determine the O_2 requirement necessary to compensate the O_2 debt accumulated during the perfusion block for the blocked tissue in kg body weight unit time.

It is easier to calculate and dose the O_2 inflow if the amount of expired CO_2 can be measured, and the amount of O_2 to be added can be set equal to it (RQ = 1). Of course, if the O_2 requirement exceeds the amount of CO_2 generated, for example, in lipid and protein metabolism, it must be satisfied, but the calculations and measurements involved are not difficult.

The following figures illustrate the possibilities for analyzing the O_2 consumption of the various organs and the whole body. O_2 consumption of the organs at rest as a percentage of the basal metabolism of the whole organism is as follows: liver, 26.4%; brain, 18.3%; heart, 9.2%; kidneys, 7.2%; skeletal muscle, 25.6%; remainder, 13.3% (this includes skin, stomach, intestines, pancreas, spleen, lungs, bone, fat, etc.). Assuming an O_2 consumption of 248.5 ml/min, as in the last example, the following amounts are calculated as metabolized: liver, 65.6 ml; brain, 45.5 ml; heart, 22.9 ml; kidneys, 17.9 ml; skeletal muscle, 63.6 ml; remainder, 33 ml. This applies for a young male patient. For the 73-year-old of our example, consumption is reduced in each case by 15% due to age. If O_2 consumption is followed from heartbeat to heartbeat, O_2 transport in the previous example can be calculated, assuming a heart rate of 62 beats/min: 248.5 ml O_2/62 beats = 4 ml O_2/beat. Thus, we may say that the O_2 pulse is 4 ml. Of this 4 ml, the following amounts are calculable to the individual organs in the following way: liver, 1.056 ml; brain, 0.732 ml; heart, 0.368 ml; kidneys, 0,288 ml; skeletal muscle, 1.024 ml; remainder, 0.532 ml. For our 73-year-old patient these values must also be reduced by approximately 15% as a correction for age, as above.

For anesthesiologists accustomed to fresh gas inflows of 2000 ml O_2/min and 4000 ml N_2O/min or even 4000 ml O_2/min and 8000 ml N_2O/min, the figures described here are alarmingly low. Taking the Avogadro or the Loschmit constant, molecular weight, number of molecules, gas density, mass, amounts, temperature, and uptake and release of O_2, N_2O, N_2, CO_2, He, etc. as a base, it is also possible to determine the number of molecules of these that are taken up and given off. In the example above, 4 ml O_2 per heartbeat was sufficient for the energy supply of basal metabolism in a 20-year-old male patient weighing 70–80 kg. In the case, each heartbeat supplies 1 kg body weight with approximately 0.0571428 ml O_2, or, in other words, each milligram body weight is supplied with 1428 million molecules of oxygen for metabolic purposes.

These figures make clear how imprecise the dosage and measurement instruments are that we have compared to the theoretical possibilities for monitoring and control of anesthesia on a molecular level. They also illustrate where developments are necessary, if we are to determine in an exclusively noninvasive manner, the diffusion processes (speed of diffusion), alveoarterial and artereovenous gas differences, stroke and minute volumes of the heart, perfusion processes, changes in dead space, and many other parameters. This would enable us to make full use of the technical possibilities of the closed-circuit system.

The above-mentioned 4 ml O_2/heartbeat, or 1428 million molecules O_2 for every milligram body weight per heartbeat, requires approximately 2.9 g hemoglobin in 100 ml plasma. In general, however, 10–16 g hemoglobin containing $3.5–5 \times 10^{11}$ erythrocytes is available in each 100 ml plasma for the transport of this 4 ml, or $2.505946 \times 4 \times 10^{16}$ molecules of O_2.

Instead of the above-mentioned O_2 pulses, one can use O_2 availability as the parameter of interest. O_2 availability is understood here as O_2 content × HMV. In our example, assuming a cardiac output/beat of 100 ml blood, O_2 availability is about 22 ml O_2, or more than five times the resting demand of 4 ml O_2. This holds for blood in the normal range, where hematocrit is between 36% and 44%, hemoglobin is 16 g %, there are 4.5–5.5 million erythrocytes/mm^3, the average arteriovenous O_2 content difference ($avDO_2$) is ca. 4.7 vol.%, and the physically dissolved O_2 in the blood is between 0.23 and 0.30 vol.%. This means that 16 g hemoglobin transport 4 ml O_2, or $2.505946 \times 4 \times 10^{16}$ molecules O_2 are transported by 5×10^{11} erythrocytes in 100 ml blood. Naturally, corrections must be made for pathological values.

These and other new parameters serve not only in the dosing of anesthetic gases and vapors but also for monitoring and diagnostic systems. The closed-circuit system has opened the way for anesthesia to advance to the molecular level, a "molecular anesthesia." All measuring and dosing instruments for anesthesia should, in future, be oriented toward this end. In the future, we will not think merely in terms of gas concentrations and partial pressures, as we learned in blood gas analysis 25 years ago, but in molecular amounts to describe consumption and gas exchange.

If nothing is lost to the external atmosphere from the closed-circuit system, we can use noble gases such as He, Ar, and Xe to improve the anesthesia and can simultaneously measure cardiovascular and pulmonary parameters without incurring significant costs.

Addition of Fresh N_2O

The use of N_2O as an analgesic in the anesthetic gas mixture has proved so useful that we would not want to dispense with it until a better analgesic gas is available. However, we will not discuss the advantages and disadvantages of the use of N_2O here.

As N_2O is not metabolized, other criteria than in the case of O_2 must be considered with respect to its uptake. Here it is not a matter of consumption, as with O_2, but rather one of uptake or substitution of an inert gas which is being given off continually by the body via intestines and skin. The desired N_2O concentration is 65–70 vol.%. The easiest way to achieve this value in the closed-circuit system, with the pressure-relief valve open, is to wash the circuit system out with a high gas flow (approximately 6 l/min) of a gas mixture already containing the desired concentration, while the N_2O concentration in the gas stream is continually measured.

The addition of gas is regulated at present using a rotameter once a minute (Fig. 2). In the future, use will surely be made of computer-controlled flowmeters with extremely fast acting valves that can close between breaths. In addition, devices are employed to measure the N_2O concentration; combined with integrated flowmeters for the measurement of tidal volume and gas concentrations, these devices make possible the exact addition of gas volumes from breath to breath.

N_2 Concentration in Anesthetic Gas Mixture

Initially, an inspiratory O_2 concentration of 25–30 vol.% and an expiratory concentration of 21–26 vol.% is aimed at. In half-open and half-closed systems an O_2 concentration of about 33 vol.% is routinely used. Since these systems are usually run with a large gas excess (generally up to 96%), the normal air mixture is quickly washed out of the lungs and replaced by the anesthetic gas mixture. Thus, within a short time there is no N_2 remaining in the lungs. Naturally, the N_2 released into the lungs from the body tissue is also quickly ventilated. In contrast, with the closed-circuit technique, some amount of N_2 remains and accumulates in the system.

By using a high fresh gas flow with the pressure-relief valve opened, the vital capacity of the lungs, with their 79 vol.% N_2, may quickly be washed out and replaced with 100 vol.% O_2 or any other gas mixture before or during the induction of anesthesia. Once the system is closed, only the N_2 released by body tissue into the lungs can collect in the system. If the lungs are not washed with pure O_2 prior to closing the system, the normal gas content of the lungs mixes with the anesthetic gases of the closed-circuit system. In either case, we have a mixture of N_2, N_2O, O_2, and other gases such as Xe and He, the concentrations of which depend on the volume of the gases (V) and the time since closing the system (*t*).

If the concentration of N_2O can be measured along with that of O_2 (Fig. 4), it is possible to determine, by subtraction, the concentration of N_2. (At the moment, N_2 concentration can only be measured by flame photometry, a technique which is not applicable to the closed-circuit technique because of the gas losses necessary for such analysis.) For example, if we measure 25 vol.% O_2 and 60 vol.% N_2O, we have, together, 85 vol.% $O_2 + N_2O$. And if the total gas amount is 100 vol.%, and we measure 85 vol.% $O_2 + N_2O$, by subtraction we arrive at a figure of 15 vol.% for other gases, e.g., N_2, CO_2, X_e, He etc.

At this point we should remember that during spontaneous respiration humans take up 79 vol.% N_2 and only 21 vol.% O_2. We have been accustomed to this proportion of N_2 in the atmosphere since the beginning of our evolutionary history millions of years ago. Thus, N_2 cannot be disadvantageous for the lungs – on the contrary, it provides protection for alveolar surfactants.

Reservations against the closed-circuit system, based on the possible increase in N_2 content of the system due to release from tissues or residual N_2 in the lungs, are groundless, both in theory and in clinical practice, as long as the criteria described for monitoring gas concentrations are met. Our own excellent clinical results have confirmed this over 70000 times. Since we can now measure N_2O concentrations along with O_2 concentrations using available instruments, we will surely soon be able to ascertain whether and how much N_2 is necessary, along with the other gases in the (closed-) circuit system, to provide protection for the surfactant factors in the lungs and to avoid the formation of shunts.

N_2O Concentration in Anesthetic Gas Mixture

Since N_2O, as well as N_2, is an inert gas, an equilibrium between its uptake and its loss, as well as between N_2O and the residual N_2 the body, is quickly established. The gases from the lungs and the anesthesia system circulating in the closed system stabilize themselves at the concentrations set. By further addition of more, less, or no N_2O together with the fresh O_2 inflow, any desired N_2O concentration in the systemic gas may be achieved. Since N_2O is constantly being lost via skin, intestines, and the rubber parts of the system, it must constantly be replaced by fresh gas inflow into the closed-circuit system.

As soon as the closed-circuit system is filled, as described above, it is important not to make any sudden, large changes in the settings of the rotameters when attempting to regulate gas concentrations. Rather, only fine changes should be made, in accordance with calculated or consumed gas volumes, even when small leaks must be compensated which cannot immediately be found.

The fineness with which the anesthetic gases and vapors can be added is limited by the following factors:

- The absolute gas tightness of the system.
- The technical perfection and precision (fine regulation) of the rotameters.
- The exactness and rapidity of the flowmeter measurements.
- The design and technical quality of the dosing and measuring devices with respect to accuracy, repeatability, and hysteresis.
- The technical reliability of the closed-circuit system and all devices associated with it.
- The comparability of the units of measurement and those of the dosing devices. This is important not only for central monitoring but also for automatic closed-loop feedback systems (e.g., O_2 uptake and CO_2 production per minute, breath, or heartbeat).

Addition of He or Other Gases

Expensive gases such as Xe and He can be economically used only in closed-circuit systems (e.g., to improve lung function or to open new avenues in the development of inhalation anesthesia). Such developments depend, essentially, upon the availability of necessary measuring instruments which are affordable, practical for clinical use, and which function without technical problems.

Since He and Xe cannot escape from a closed-circuit system, they need only be added in small amounts, and the costs, even for such expensive gases, are therefore not prohibitive. The costs may be further reduced because most of these gases may be recovered during the termination of anesthesia.

The Apparatus

Rotameter for Dosing of Gases

The addition of the anesthetic gases O_2 and N_2O is accomplished using specially produced rotameters made by Dräger. The units are graduated as follows: from 20 to 500 ml/min in 10 ml/min steps; from 500 to 1000 ml/min in 50 ml/min steps; from 1000 to 1500 ml/min in 100 ml/min steps; from 1.5 to 5 l/min in 500 ml steps; and from 5 to 10 l/min in 1 l/min steps (see Fig. 2).

Disadvantages of these devices include the following:

1. (a) The dosing steps of 10 ml/min are still too imprecise for our purposes and, in addition, are not accurate. The floater for this control is extremely light and its position fluctuates with the changes in intrasystemic pressure which occur with every breath. (b) It is not possible to dose with a fineness of 1 ml/min. An even lighter floater which would be necessary for this and would, of course, be even more sensitive to changes in volume and pressure.
2. (a) The fine adjustment of the different gases with this device requires time, patience, and much manual dexterity. (b) The time taken to adjust the gas volumes leads to further inaccuracies in their measurement.

Characteristics of an ideal device would include:

1. (a) Gas dosing units which open and close quickly, and which are computer driven. (b) Precise measuring devices (with an accuracy of at least 1 ml/min) for the exact measurement of inspiratory and expiratory volumes of the anesthetic gases and vapors, such as O_2, CO_2, N_2O, He, Xe, halothane, and H_2O, which are clinically usable. This should be the next step in an interim solution.
2. As a final solution for routine clinical use, precise measuring devices should be developed which have a measurement accuracy in the order of 10^6 molecules O_2 (per milligram body weight and per heartbeat) and are integrated into an electronic data processing system.

Devices for Measuring O_2 Concentration

Relative concentrations of the gases O_2 and N_2O may be determined by measuring only the O_2 concentration (vol.%) of the anesthetic mixture. For this purpose we can use the following:

- The Oxycom device (Dräger), which is based on an electrochemical principle (Fig. 3).
- The Normocap made by Datex use a polarographic principle (Fig. 4).

– The Multicap or Capnomac (Datex), which measures oxygen paramagnetically (Fig. 4).

For reasons of safety (because of the small fresh gas flows and the relatively low O_2 concentration) we measure continually in both the inspiratory and expiratory parts of the system.

The major disadvantage of the Dräger devices lies in the relatively long time required to determine a measurement (up to 18 s). Only the paramagnetic measurement of O_2 is much quicker (0.1 s). Ideally the speed of measurement would be in the order of milliseconds.

Devices for Measuring N_2O Concentration

Recently, units such as the multicap or Capnomac (Datex) have become available which are capable of measuring N_2O together with N_2, CO_2, O_2, and volatile anesthetics, even in the presence of high concentrations of water vapor, as is the case in the closed-circuit system (Fig. 4). Thus it is now possible to determine *indirectly* (by exclusion) the N_2 concentrations in the closed circuit, as well as to determine directly the concentrations of O_2, CO_2, and volatile anesthetics. Nevertheless, the trace concentrations of inert gases (He, Xe, etc.) remain neglected. With the help of these N_2O measurements together with the O_2 measurements) we can calculate the amount of N_2 in the system at any time and reduce it to any desired level by opening the pressure-relief valve and washing the closed-circuit system with a larger gas flow (e.g., 1.5 l O_2 and 4.5 l N_2O per minute).

The advantage of this unit is that gas samples are measured without changing them physically or chemically, so that they may be returned to the closed system for reuse. In contrast, mass spectrographic measurements would require the intermittent removal of larger quantities of gas from the system, which are then chemically transformed by the process of the analysis. For this reason mass spectrometry is not technically relevant for use with the closed-circuit system, even if it were not prohibitively expensive.

Devices for Measuring N_2 Concentration

At present N_2 can only be measured by flame photometry. For each such measurement, a volume of gas (approximately 5 ml) must be removed, which cannot be returned to the system afterwards due to the chemical changes caused by the measurement procedure. Furthermore, it is not possible with this technique to make continuous, or even frequent measurements. Thus a problem-free and clinically practical method for the direct measurement of N_2 concentration in the anesthetic gases is not available.

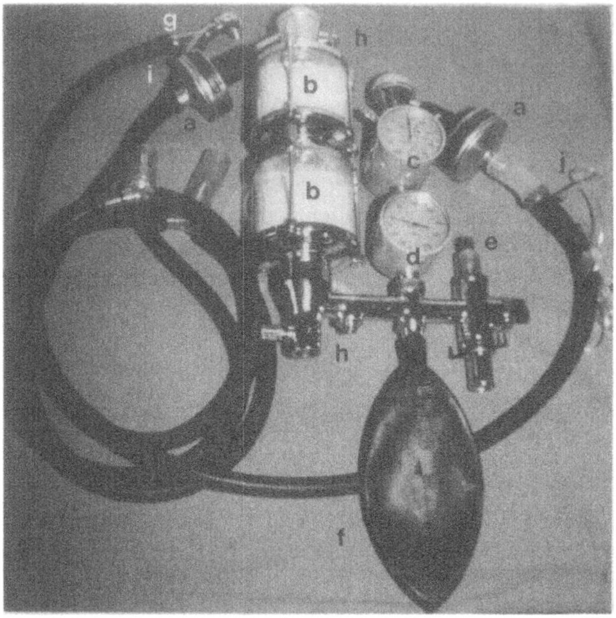

Fig. 6. Closed-circuit systems. *a,* Bacteria filter (in inspiratory and expiratory branches); *b,* CO_2 absorber units; *c,* volumeter (Dräger) for measuring tidal volume; *d,* pressure gauge (Dräger) for measuring the pressure in the breathing system; *e,* pressure-relief valve (pop-off valve); *f,* reservoir bag; *g,* Y piece; *h,* fresh gas inlet; *i,* suction tube for gas sampling; *j,* tube for returning gas samples from the measurement instruments to the closed-circuit system

In the future, the measurement of concentration and volume of N_2 in the anesthetic gas mixture, along with those of N_2O, CO_2, O_2, the noble gases, volatile anesthetics, and H_2O should be possible, quickly and accurately, from breath to breath and without chemical or physical changes in the samples.

The Closed-Circuit System

We use the commercially available closed-circuit system made by Dräger (Fig. 6). For use in a closed system, specially selected, high-quality parts must be used which fit together perfectly. Even a leak of a few milliliters per minute is not acceptable if the advantages of the closed system over half-open and half-closed systems are to be fully exploited. The matched pieces of a leak-free closed-circuit system must be supported and maintained with the utmost care.

The systems used for half-closed operation generally have leakages of several hundred milliliters per minute and up to 1000 ml/min or more. Systems of such low quality are totally useless for closed-circuit operation and are actually intended for nonclosed operation. Through these leakages unacceptably large amounts of anesthetic gases and vapors are given off into the air of the operating room. The conventional systems for air filtering and "pipeline systems" do not reduce this problem at all, as they remove only the gases escaping from the pressure-relief valve and not those gases and vapors escaping before the valve along the entire length of the system. Some parts of a leaky system can be so improved as to be useful. Where this is not possible, the parts must be replaced by others selected by the manufacturer for high quality.

For over 15 years, manufacturers have been urged either to assemble systems from much more precisely produced pieces (also for nonclosed systems) or at least specially to finish the pieces used for closed systems. Even better would be the implementation of a totally new design; this is now long overdue.

Disadvantages of presently available systems involve the following points: (a) the volume of circulating gas in the system is too large; (b) there are too many connecting parts; and (c) the connections between the individual parts are too imprecisely finished and matched. The connections between tubes and metal parts are still made with particularly little care and precision.

Characteristics of an ideal system would take into consideration:

1. Minimization of circulating gas volumes.
2. Reduction in the number of connecting parts.
3. (a) Specially designed connections which are absolutely gas tight, do not slip, and make secure durable contacts. (b) The gas volume of the system, including the individual parts, should be measured exactly by the manufacturer and given on the parts. (c) Only ultralight materials should be used in the manufacture of the systems, but at the same time they should be robust enough to withstand mechanical cleaning, disinfection, sterilization, disassembly, assembly, and the full spectrum of clinical uses without damage.
4. Technically sophisticated, standardized coupling parts for the connection of measuring devices.

Breathing Tubes and Water Traps

It is imperative that breathing tubes have smooth inner walls and good thermal isolation characteristics. As the anesthetic gases cannot escape from the closed system, the circulating gases quickly become saturated with water vapor. This is of great advantage to the patient, as we have previously confirmed. However, as soon as these water vapor saturated gases reach the expiratory side of the system and undergo cooling from body temperature to room temperature, part of the water vapor condenses, and water droplets form on the inner wall of the breathing tube and at measurement points. These water droplets disturb the quantitative measurement of CO_2, N_2O, N_2, O_2, noble gases, and volatile anesthetics.

Tubes with smooth inner walls allow condensed water to collect at the lowest point, where it can be collected in a container, as so-called water trap (see Fig. 7). In folded tubes the water cannot run down, but collects in the folds. Once this has occurred, the tubes can only be dried by stretching and warming while being washed with dry gases over longer durations. A water trap does not help with folded tubes. There is no alternative to breathing tubes with smooth inner walls that are thermally, poor conductors when the closed-circuit system is used.

Disadvantages of the breathing tubes and water traps include the following:

1. (a) Every additional part introduces two possible sources for a leak. (b) Connections are not adequately firm and may be disconnected inadvertently.

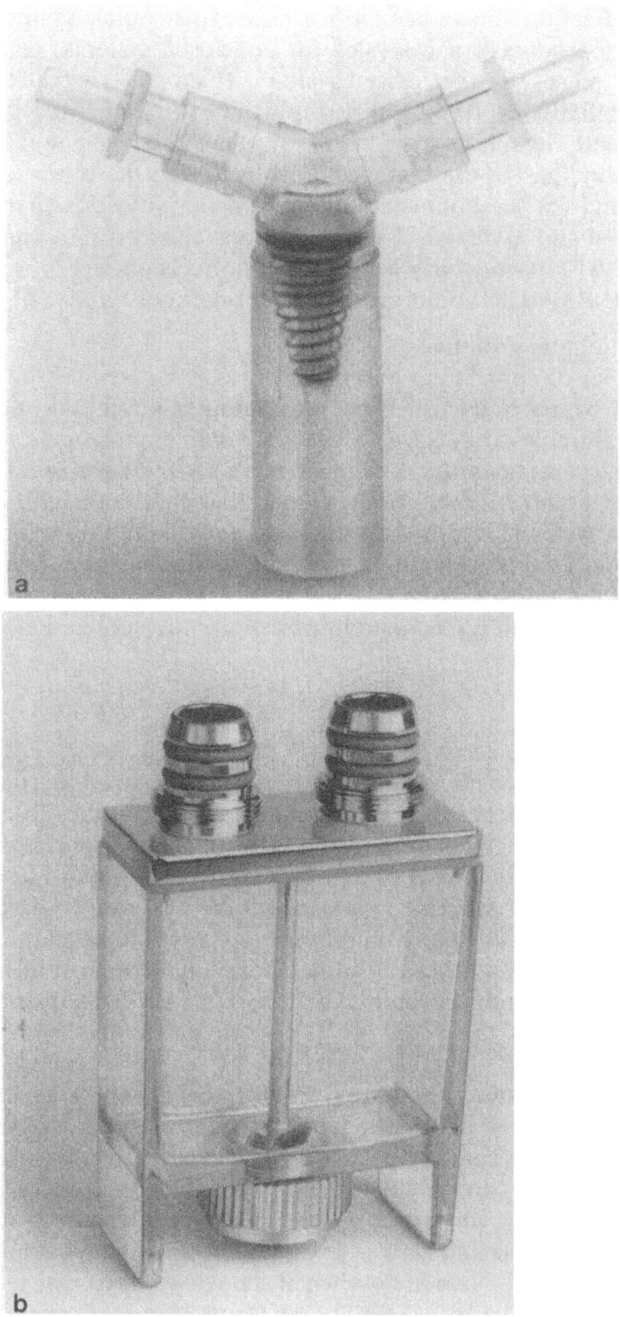

Fig. 7a, b. Various types of water traps (Dräger)

2. Tubes are too heat conductive.
3. Tubes are not absolutely gas tight.
4. Tubes are much too heavy.
5. It is difficult and awkward to connect the breathing tubes to the patient.

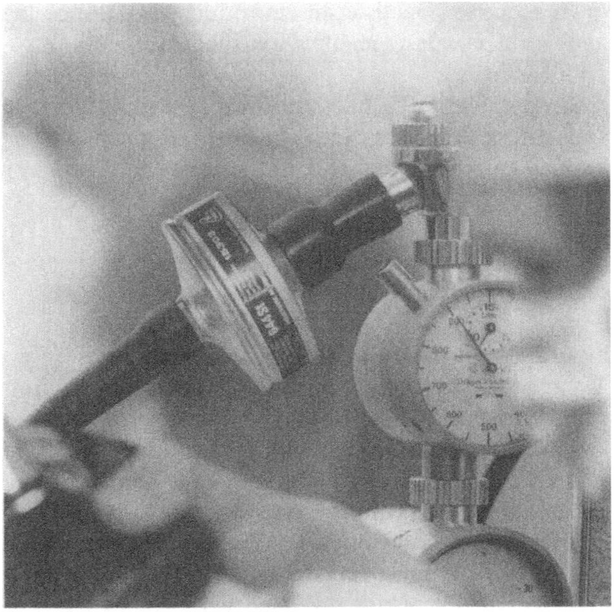

Fig. 8. Microbe filters (Dräger) for insertion in the inspiratory and th expiratory branches

6. Gas volumes of tubes and water traps are not specified by the manufacturer.
7. Breathing tubes are available in only two lengths.
8. The lowest point, where the water trap should be connected varies.

Characteristics of an ideal system would include:

1. Absolutely gas-tight and automatically sealing connectors, usable between all components.
2. Effective thermal isolation.
3. Absolute gas tightness.
4. Easily readable specifications for gas volume on all parts.
5. Tubes with smooth inner walls, variable in length.
6. Ultralight tubes.
7. A practical design for self-sealing of tubes before and during the operation and in storage after the operation.
8. A fixed lowest point for tubes.

Bacteria Filter

In order to save the time and work of cleaning and sterilization, it is recommended that bacteria filters (Fig. 8) be built into the inspiratory and expiratory parts of the closed-circuit system. This also avoids contamination of the measurement devices which via bypass analyze gas samples (e.g., CO_2, N_2O, O_2, volatile anesthetics) and afterwards return them to the closed-circuit system. After the measurement, in every case the gases pass through filters before being returned to the patient.

In addition, bacteria filters reduce turbulence in the system, which is important for accurate concentration measurements (e.g., with the Emma units made by Engström).

In summary, the bacteria filters not only keep the closed-circuit system and measuring devices reliable and hygienic, they also promote stable functioning of the whole system and reduce the time and costs necessary for keeping it clean.

Supply of Volatile Anesthetics

For the vaporization of volatile anesthetics (e.g., halothane) we prefer, as discussed above, the Vapor (Dräger). We use units which can develop an end concentration of 7 vol.% In fact, two normal vaporizers, which, for example, develop a maximum of 4 vol.% halothane each, may be used in series, enabling an end concentration of 8 vol.%. Low gas flows can only take up small amounts of the anesthetic vapors from vaporizers. This may be compensated by increasing the vapor concentration of the vaporizer. When lower gas flows are used, the vaporizers must have a higher upper limit for releasing volatile anesthetics than vaporizers which are used with higher fresh gas flows.

For example, a fresh gas flow of 2000 ml O_2/min and 4000 ml N_2O/min takes up 30 volume units (depending on temperature) of volatile anesthetic per minute when the vaporizer is set at 0.5 vol.%. If the fresh gas flow is reduced to 245 ml O_2/min and 105 ml N_2O/min, only 1.75 volume units/min are delivered to the patient. With the small gas flows used in closed-circuit systems, 30 volume units/min volatile anesthetic can only be transported to the patient when the vaporizer setting is correspondingly increased. With a vaporizer setting of 7 vol.% and a total fresh gas flow of 350 ml/min, we achieve an anesthetic vapor transport of 24.5 volume units/min. With two normal vaporizers in series, developing an end concentration of 8 vol.%, a transport of even 28 volume units is possibe (naturally, these values are all temperature dependent). As this example shows, the increased vapor concentration of volatile anesthetics flowing into the closed-circuit system is necessary only in order to maintain a sufficient amount of (volatile) anesthetic/min for the patient.

The closed-circuit system gives information, not only about the amounts of O_2 and N_2O taken up per minute, but also about the uptake of volatile anesthetics. The closed-circuit system is, thus, not only an anesthetic system but also a monitoring system for the amounts of anesthetic gases and vapors taken up and given off by the patient. This twofold function, as anesthetic and as monitoring system, brings a further advantage over other anesthesia systems.

Since excessive amounts of anesthetic gases are not introduced into the closed-circuit system, the concentration of anesthetic vapor can only rise correspondingly slowly from the expiratory to the inspiratory segments of the system. The exactly dosed addition is only possible with highly efficient vaporizers and continual monitoring of the inspiratory and/or (end-)expiratory concentrations from breath to breath. The higher the amount which returns from the patient through the expiratory branch, the less fresh gas may be added to the system on the inspiratory side, and vice versa.

If the necessary end-expiratory halothane concentration cannot be reached or maintained with low gas flow because vaporizers enabling such a concentration at this low flow are not available, one may switch over to half-closed operation with a more or less openend vaporizer valve for a short time (usually only necessary for a few breaths) until the necessary concentrations and amounts of volatile anesthetics are achieved in the system. This changeover of the mode of operation is usually necessary for only a few breaths at correspondingly higher gas flows, but it is only a makeshift remedy, as in this time all the advantages of the closed-circuit system are lost.

It is the amount of volatile anesthetic taken up by the patient and not the amount used by the system which remains the same between open and closed operation of a system. As already mentioned, it is because of the amount to be taken up that the concentration of the volatile anesthetic in the closed system must be higher at the beginning of anesthesia than in a half-closed system. Over time, as the concentration of the anesthetic in the patient's body builds up and approaches the steady state, the amount of fresh anesthetic needed becomes less and less. This may be seen in the fact that the difference between the inspiratory and (end-) expiratory concentration of the anesthetic measured decreases. Low attempted to describe this fact in his "square root of time model:"

$$\text{Rate uptake at t minutes} = \frac{\text{rate uptake at 1 minute}}{\sqrt{\text{t}}}$$

However, the use of this formula cannot adequately replace, either physiologically or pharmacologically, the empirical determination of the anesthesia requirement based on appropriate measurements.

The characteristic differences between the closed-circuit system and the other more or less open systems lie in the magnitude of the total gas flow per minute, the resulting amount of anesthetic available, and the information available regarding the uptake, release, and consumption of anesthetic in each time interval. As may be seen in the previous example, the smaller the gas flow is, the higher the vapor concentration developed by the vaporizer must be, and vice versa, in order to achieve clinically sufficient levels of anesthetic in the plasma and tissue.

Measurement of the Concentration of Volatile Anesthetics

The concentrations of the volatile anesthetics may now be sufficiently accurately measured using devices available from a number of companies (Engström, Datex Fig. 4; Dräger, Siemens; Fig. 9), which use the piezoelectric or infrared absorption principle. These are quite useful for clinical applications.

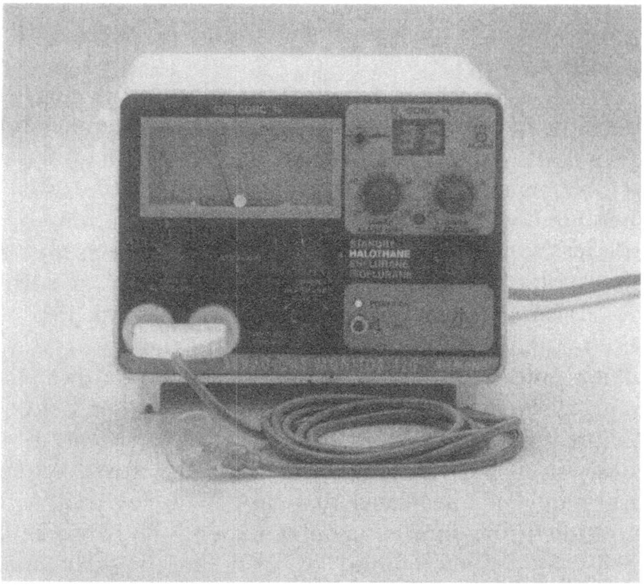

Fig. 9. Servo-Gas-Monitor 120 (Siemens) for determining concentrations of volatile anesthetics

As a rule, an end-expiratory concentration of 0.5–0.6 vol.% halothane is sufficient for an elderly patient and 0.7–0.9 vol.% for a young patient in good physical condition. Naturally, clinical experience must also here be a guide when anticipating the anesthetic requirement for a patient, taking into account the planned operation, the constitution, sex, and health of the patient, as well as many other factors.

The measurement accuracy of these units varies from manufacturer to manufacturer due to the varied influence of the water vapor saturation of the gases in the system; the magnitude of measurement errors also varies occasionally between units of the same manufacturer. For this reason, the measurement instruments, as well as all other parts of the closed-circuit system, should not be exchanged between systems, but kept together as a single integral unit. For clinical work, the absolute value of the volatile anesthetic concentration is not as important as reliable relative measurements, since the effect of the anesthetic varies so strongly from patient to patient and must, in any case, be adapted to the clinical requirement.

As in nonclosed systems the various brands of halothane (halothane, Halothan-Halocarbon, and Fluothane) exhibit no differences in clinical effect in the closed-circuit system. Halothan-Halocarbon is the least expensive of these. We have also failed to find any clinically relevant differences between the various brands of halothane and isoflurane. The future will show to what extent isoflurane, and its metabolites, is superior with respect to hepatic stress, brain perfusion, and, in the final analysis, for the whole organism. Even the recovery phase is not superior with isoflurane when the closed system is used with continual monitoring and control of the inspiratory and end-expiratory concentrations. The significantly higher price of isoflurane, coupled with the higher doses required and the uncertainty with respect to side effects gives us no reason to switch over to this anesthetic at the moment, even though its use in the closed system is much less expensive than in nonclosed systems.

The possibilities for exact control of the anesthesia provided by the closed-circuit system and its integrated measuring instruments restore halothane to the position it deserves on the basis of its pharmacological qualities.

A disadvantage of concentration-measuring devices is the sensitivity of units to water vapor, and this can cause significant variation from the absolute value. For clinical work this is relatively unimportant, as long as the instruments are not exchanged with in a system. Such changes always require reorientation, which is distractive and irritating. A further disadvantage of present instruments is that it is still not yet possible to measure the amounts of anesthetics taken up by the body per breath or per minute. Exact measurements of the tidal volume and of the difference between inspiratory and expiratory concentration would allow calculation of the amount of volatile anesthetics taken up per breath or per minute.

In an ideal system, the units would be able to measure the concentrations and amounts of gases and volatile anesthetics taken up by the patient quickly (in the order of milliseconds) and give absolute measurements at any point in the breath cycle. They would also be able to measure He and Xe in the same manner. Along with the gases and volatile anesthetic, the units would, in addition, be able accurately to mesure the H_2O concentration (degree of water vapor saturation).

Respiration

All conventional respirators exhibit significant leakages. When such respirators are used with a closed-circuit system, the danger of the system being "breathed up" is great. Until the closed-circuit system becomes available, manual respiration must be used.

If a coupling system in the form of a "bag-in-box" system (Fig. 10) or a "rolling seal" system (Fig. 11) is introduced between the gas-tight closed-circuit system and the respirator, so that the mechanical respiration is driven by the respirator, it is possible to use a conventional respirator without risk. In this case it is of particular importance that the coupling system (box) is made of a transparent and absolutely gas-tight material (e.g., Plexiglas). The filling condition of the closed-circuit system may thereby be monitored at any time, not only by use of intrasystemic pressure measurements, but also visually, by observing the reservoir bag. Figures 10 and 11 show the bag-in-box system and the rolling-seal system as well as their combination (Fig. 20) with the closed-circuit system and a respirator (Servo-ventilator, Siemens).

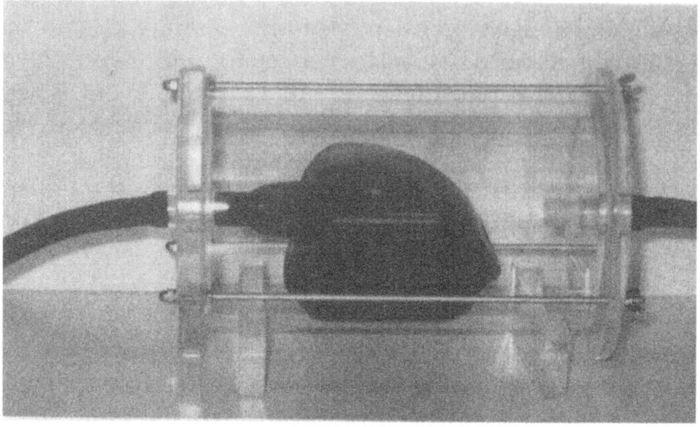

Fig. 10. The "bag in box" system. This system supplants conventional mechanical respiration via reservoir bag. Instead of compressing the bag by hand, it is enclosed within a transparent cylinder and its compression is pressure driven from a conventional respirator. In order to save costs, the respirator is run with compressed air instead of pure O_2. With this construction it is irrelevant whether the respirator itself is gas tight, or whether it was designed for nonclosed systems such as the Servoventilator 900 C (Siemens). The cylinder enclosing the reservoir bag must be so constructed that it can be opened in order to clean and replace the bag after use. The reservoir bag must be gas tight to prevent gases from the closed-circuit system leaking into the cylinder. Conventional reservoir bags fulfill this requirement. The cylinder walls may be made of any of a number of transparent materials, in order that the filling status of the reservoir bag may be directly observed. Thus, the filling of the closed-circuit system may be checked at any time, not only by pressure measurements, but also by direct visual monitoring of the bag. This is an important prerequisite for changing over from hand- to machine-driven respiration, if unintentional breathing up of the closed-circuit system is to be ruled out. The bag-in-box system, as presented here, with a transparent cylinder for coupling respirator and closed-circuit system, is ideal. (This bag-in-box system was built for us by B. Lachmann, Erasmus University, Rotterdam, Holland)

Control of Respiration

When CO_2 concentration is measured breath by breath and the end-expiratory value is kept between 4.0 and 5.6 vol.% with adequate tidal volume, we speak of normoventilation. We attempt to hold the end-expiratory CO_2 concentration at 4.8 vol.%, that is, at the midpoint of the normoventilation range of 4.0–5.6 vol.%. Comparison of simultaneous blood gas measurements has repeatedly convinced us of the appropriateness of this strategy, and we have now adopted it. An undershooting or overshooting of the normoventilation boundary condition should be strictly avoided. We adopted monitoring and control of respiration according to the end-expiratory concentration of CO_2 as standard over 15 years ago in order to avoid, on the one hand, hyperventilation-induced inadequacies in brain circulation due to reduced pCO_2 and, on the other hand, hypoventilation-induced increases in intracranial pressure due to increased pCO_2.

For regulation of respiration, the tidal volume together with the end-expiratory CO_2 is crucial. In order to avoid, a priori, such problems as hypoventilation due to increase in dead space and disturbances of perfusion, diffusion, and

Fig. 11. The rolling-seal system. Gas tightness is achieved in this system by use of a rubber gasket mounted over the piston between it and the inner wall of the cylinder. This rubber membrane is stretched over the piston in such a way that it does not interfere with the movement of the piston within the cylinder. By the use of this rather expensive technique it is possible to maintain system sealing, even in the area of the tidal piston. A significant disadvantage of the rolling-seal system at the coupling between the respirator and the closed-circuit system is that leaks here are not "visible" and can only be detected with reference to intrasystemic pressure measurements. A further important disadvantage of this system is that the piston gasket must be highly elastic and must, for this reason, still be made of rubber. The constant use of volatile anesthetics such as halothane quickly results in deterioration of the gasket material and the material of the cylinder wall (For the construction of the rolling-seal system we thank the company, Mijnhardt, Netherlands)

metabolism, we use a tidal volume between 600 and 800 ml per breath for adults of normal weight who present no conspicuous respiratory symptoms. In addition, normoventilation is monitored in the closed-circuit system in terms of the following parameters: (a) O_2 uptake; (b) CO_2 release in milliliters per minute, CO_2 release in milliliters per breath, and end-expiratory CO_2 (vol.%); (c) the dead space; and (d) the alveolar ventilation. Use of the closed-circuit system, combined with measurements of O_2 concentration (vol.%) and consumption (ml) and end-expiratory CO_2 (vol.%) and CO_2 amount (ml), makes it possible to keep respiration easily within the limits of normoventilation. Falls in blood pressure due to hyperventilation and rises in blood pressure due to hypoventilation can be recognized clearly at the outset and thus avoided. Changes in pH (e.g., due to anaerobic metabolism following partial interruption of the circulation) can also be immediately recognized with this method and can be quickly corrected by compensating end-expiratory CO_2. The same is true for electrolyte changes, which can be induced by deviations from normoventilation.

Halothane anesthesia-induced decreases in blood pressure may be distinguished from hyperventilation-induced decreases when the end-expiratory concentrations of both the volatile anesthetic and CO_2 are measured and both are accordingly compensated. The end-expiratory CO_2 is, as we know, a very sensitive parameter for the circulation. In the extreme case of cardiac arrest, O_2 transport ceases, the O_2 pulse sinks to zero, and there is a parallel fall to zero in CO_2 release. In this case, the decline in end-expiratory CO_2 is not an indication of hyperventilation but rather a sign of circulatory failure (decrease in blood pressure). Thus the closed-circuit system gives warning of cardiovascular arrest through the O_2 uptake and CO_2 release values, when it is noted that intrasystemic pressure increases due to continued fresh gas addition because (a) O_2 is no longer taken up or (b) CO_2 is not released via the lungs.

The closed-circuit system is a monitoring system because, for example, it immediately indicates the cessation of O_2 and CO_2 gas exchange. It is, of course, also possible in all nonclosed systems to determine CO_2 values, but the simultaneous check by measuring O_2 uptake or O_2 consumption is not yet possible.

At present, the simultaneous determination of CO_2 release and O_2 uptake is indispensable for judging the effectiveness of heart massage during reanimation attempts. As a measurement and control parameter, end-expiratory CO_2 is not only sufficient, it is far superior to ECG recording. Among other reasons, this is because during the reanimation attempt it may be measured continually from breath to breath without any problem and gives an, immediate measure of circulation, rather than the indirect information of the ECG obtained from changes in summation vectors related primarily to the bioelectrical processes of the heart. O_2 uptake and the relationship between CO_2 release and O_2 uptake gives the same information.

What has been described here for the extreme case of cardiovascular arrest is of course also true for the less extreme cases of decrease in blood pressure. One must remember, however, that hyperventilation-induced hypocapnia also leads to reduction of blood pressure and brain perfusion. Normoventilation (normocapnia) avoids this type of blood pressure decrease and O_2 insufficiency in the brain.

Measurement of CO_2 Concentration

We use the measuring instruments made by Siemens, Engström, and Datex, which work on the principle of infrared absorption (Fig. 4, 12). In the closed-circuit system, the water vapor saturation of circulating gases disturbs the measurements of these instruments. Accurate measurements may be achieved with these instruments in the main stream and in the bypass if (a) water traps are used at the lowest points of the expiratory and inspiratory part; (b) breathing tubes with smooth inner walls (no corrugated tubes) are used; and (c) water traps are used immediately prior to the inlet to the measurement chamber of the CO_2 measuring device.

For the determination of CO_2 concentration, the measurement chamber must be heated in order to avoid the formation of interfering water condensation. The instruments made by Dräger (Capnolog) and Siemens (Analyser 930; Fig. 12) are installed with the measurement chamber in the main gas stream, directly after the patient's mouth. The disadvantage of this technique is that the

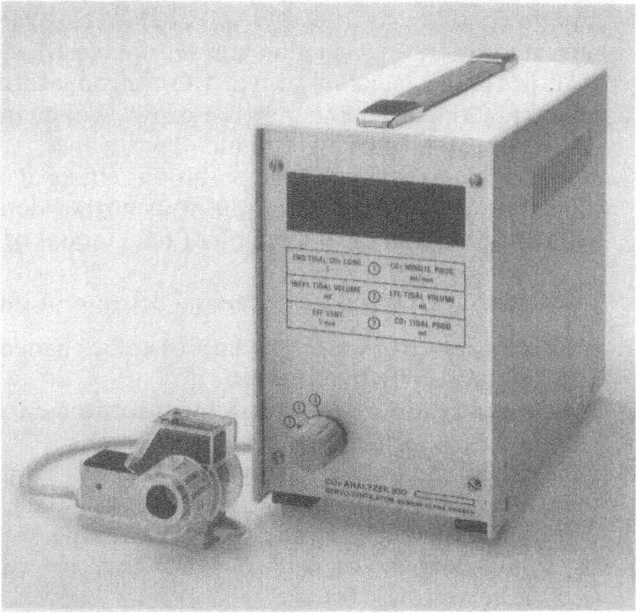

Fig. 12. Analyser 930 (Siemens) with measurement transducer that we use for the determination of alveolar ventilation per breath (using effective tidal volume in ml, and effective ventilation per minute in l/min), dead speace per breath (using ineffective tidal volume in ml), CO_2 volume per breath (using CO_2 tidal production in ml) and CO_2 volume per minute (using CO_2 minute production in ml) and end-expiratory CO_2 concentration per breath (using end tidal CO_2 concentration in vol.%) directly in the main gas stream

measurement chamber occasionally becomes too hot, and if the endotracheal tube kinks under the weight of the heated chamber, the chamber can come to lie on the patient's skin or mucous membranes and cause burns.

A disadvantage of measuring in the main gas stream it that the sensor is still to heavy and unwieldy. The advantage, on the other hand, is that measurement of CO_2 is made breath by breath and without time delay. Disadvantages of measuring in the bypass include: (a) time delay of measurement due to path length of the bypass, (b) leakages in the inlet and return paths, (c) blockages of the inlet path, and (d) suction pump failure. The fundamental advantages of CO_2 measurements, either in the main or bypass gas stream, are the following:

1. All the possibilities are given for exact manipulation of respiration (normoventilation (normocapnia) to hyperventilation (hypocapnia)).
2. An improper intubation may be immediately recognized (gas exchange does not occur in the esophagus, and a CO_2 reading is therefore not possible).
3. An indication of the metabolic situation is given by the RQ. (a) For example, RQ = 1 for carbohydrate metabolism, as O_2 consumption is equal to CO_2 production. In purely fat metabolism the RQ is 0.7. In this case, part of the O_2 uptake goes for the oxidationof H_2 to H_2O so that less CO_2 is produced for the amount of O_2 taken up than in carbohydrate metabolism,

and, in consequence, the RQ sinks below 1. (b) Pathological metabolism is indicated by CO_2 production and O_2 uptake when they are higher or lower than in resting metabolism. (c) CO_2 introduced by other sources, such as endoscopy (laparoscopy or arteroscopy), will be indicated if it gets free and causes threatening changes in blood gas values.

4. It is possible to estimate lung perfusion, either in simplified form as a graph of expiratory CO_2 concentration or in further detail as the amount of CO_2 expired per breath, as a function of the amount of O_2 taken up per breath.

A decrease in CO_2 can be registered because of the following:

1. Measurement error (check by lack of color change of the soda lime, lack of heat production by the soda lime).
2. Condensation of water vapor in the measurement chamber.
3. Disconnection of the measuring instrument.
4. Diminished O_2 transport subsequent to (a) cardiovascular failure or (b) reduction of tissue perfusion due to shock, circulation block, etc.
5. Diminished metabolism (anesthesia, poisoning, dysfunction of the hypophysis, hypothalamus, adrenalgladns, and thyroid); (b) diminished muscle tone (muscle relaxation), (c) age.
6. Hypothermia.
7. Hyperventilation.

An increase in CO_2 can be registered because of the following:

1. The CO_2-absorbing soda lime is exhausted.
2. CO_2 is introduced into the patient's body, for example, during laparoscopy.
3. Metabolic disturbances (hyperthyroidism, hyperthermia, functional disturbances of the hypophysis, hypothalamus, adrenalglands).
4. Blocking off of a region of body tissue from O_2 supply for a time, followed by its reopening to the circulation (e.g., when a circulatory block of an extremity is lifted or an arrested heart is restarted).
5. Hypoventilation.
6. Measurement errors.

Measurement of Tidal Volume

For the determination of tidal volume we use the mechanical volumeter (Dräger; Fig. 14) and the flowmeters of the respirators (e.g., Servoventilator, Siemens; Figs. 13, 15). The disadvantages of the mechanical Volumeter are: (a) large measurement errors, (b) inaccurate measurement scale with units of 50 ml, and (c) lack of electrical registration of measurements for further analysis. In contrast, the flowmeters designed according to the principles established by Fleisch, which are built into the Servoventilator, offer great accuracy and the possibility of eletronic recording of the measurements for computer-assisted data processing. Such recordings are necessary, for example, for the determination of gas amounts, gas transport curves, dead space, alveolar venti-

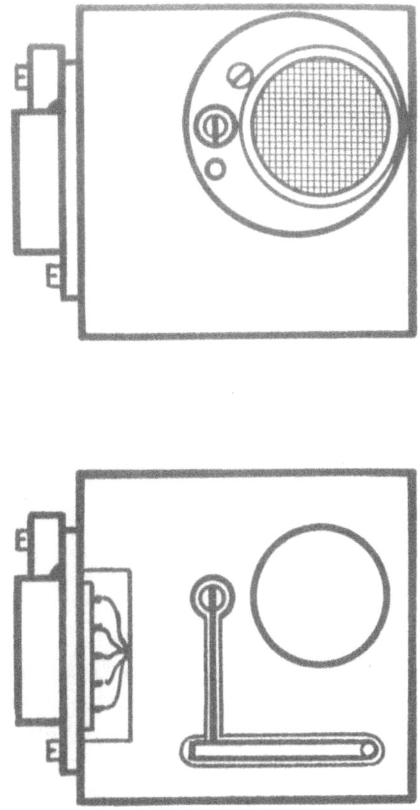

Fig. 13. Front and rear views of the flow converter of the Servoventilator (900 C Siemens) for measuring the inspiratory and expiratory gas streams

lation, RQ, O_2 pulse, O_2 availability, O_2 tissue supply, O_2 debt, disturbances of perfusion and diffusion, cardiocirculation, and metabolism, control of the anesthetic gas dosing units, cardiac output, respiration pattern, respiration rhythm, etc.

Perfect flowmeters allow exact and rapid determination of tidal volume and the determination of the amounts of O_2, CO_2, N_2O, N_2, volatile anesthetics, and noble gases taken up and given off by the patient, as well as many other parameters when correlations with the gas and vapor concentrations are made. For these purposes, it is necessary to have exact and continuous measurements of the tidal volume and the concentrations of gases and vapors over the whole period. In this way it is possible to determine these amounts even at the molecular level per heartbeat.

Determination of CO_2 Volume. The product of the tidal volume and the difference between inspiratory and expiratory concentrations of CO_2 is the basis for calculating the amount of CO_2 produced (in milliliters or number of mulecules) per breath or heartbeat (Fig. 12).

Fig. 14. Volumeter (Dräger) for measuring tidal volumes

Determination of Alveolar Ventilation and Dead Space. Computer-assisted calculations upon measurement data such as tidal volume and inspiratory and expiratory CO_2 and O_2 concentrations allow, among other things, a running determination of alveolar ventilation and dead space. Formation of a pulmonary embolism, for example, shows up immediately as an increase in the dead space. This continual determination of alveolar ventilation and dead space is of great importance for the recognition of disturbances in diffusion and perfusion, as well as for determining the magnitude of obstructive ventilation disturbances in the lungs. This is all possible and controllable in principle, with builtin redundancy, by the use of closed-circuit system as a monitoring system.

Measurement of H_2O Vapor Concentrations. Everything collects in the closed-circuit system, including water vapor. The possibility of measuring water vapor concentration and amounts would be a desirable development for the future, as such measurements are at present necessary for the accurate determination of the partial pressures, concentrations, and volumes of the other gases and vapors in order to avoid measurement errors and to make absolute measurements possible (Fig. 17). We should not be satisfied with orienting ourselves to the condensate in the water traps as an indication of water vapor saturation. That the volatile anesthetic can be measured, and that, as we know, such measurements are affected by the amount of water vapor, strongly indicate the necessity of measuring H_2O concentrations (vol.%) and volumes (ml) in the breathing mixture from breath to breath.

Metabolites of Anesthetics. Along with the anesthetic gases, the CO_2-absorbing soda lime (weight and chemical changes), water, and volatile anesthetics, the

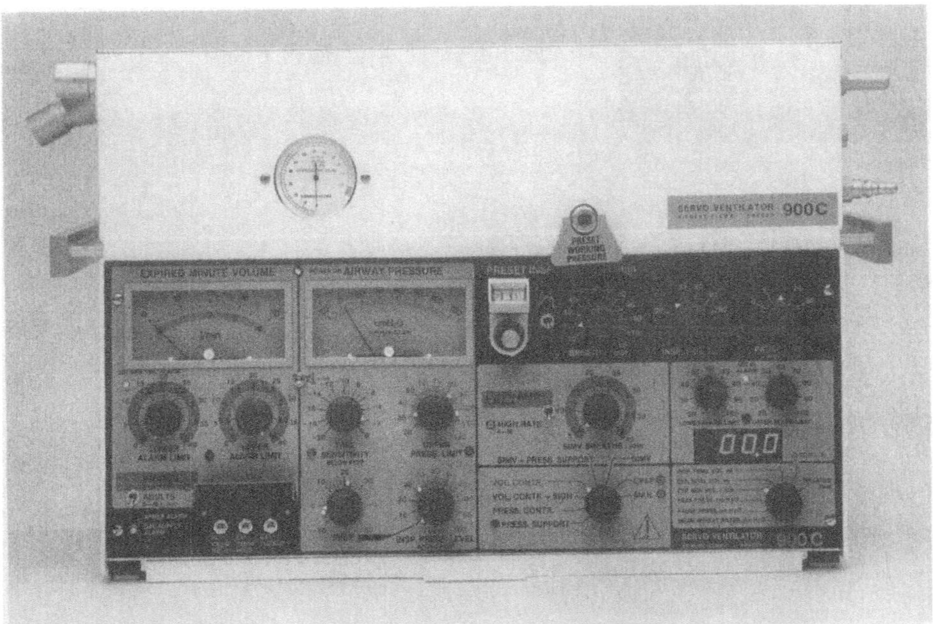

Fig. 15. The Servoventilator (Siemens) is a respirator designed for use with open systems. However, in conjunction with the bag-in-box system, or rolling-sealsystem it can be used with fully closed-circuit systems, as the system, on its side of the bag-in-box, remains fully closed within itself. The respirator is only indirectly connected to the closed-circuit system in that the reservoir bag or rolling seal, in a closed container (box), is compressed by the pressurized air coming from the respirator (see also Figs. 10, 11)

metabolites of these and other substances collect in the system and may be analyzed, quantitatively and qualitatively, in the service of pharmacokinetic ends.

Respiratory Frequency

The respiratory frequency that is set is determined by the boundary conditions for normoventilation (normocapnia) described above: tidal volume of 600–800 ml and end-expiratory CO_2 of 4.8 vol.%, whereby end-expiratory CO_2 is a fixed value and tidal volume varies from infant to adult, being determined by the pulmonary volume of the patient. Thus, we do not control respiration by setting fixed values for tidal volume and frequency but rather by keeping tidal volume and end-expiratory CO_2 fixed. Of course, when ventilation problems are present, tidal volume and respiratory pressure must be adjusted accordingly. With this technique then, respiratory frequency is only a resultant parameter which, for calculation purposes, remains as significant as the minute volume.

Fig. 16. Pressure gauges (Dräger) for pressure measurements in the closed-circuit system

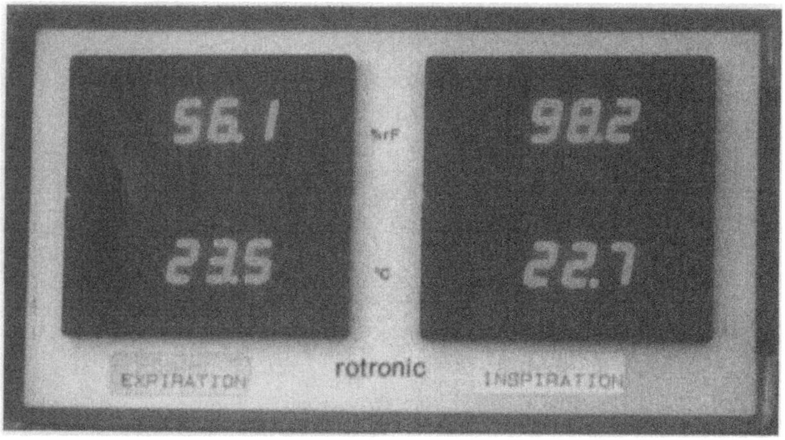

Fig. 17. Equipment for measuring humidity and temperature (Rotronic)

Respiratory Pressures

The determination of respiratory pressures in the closed system is made as in half-closed and half-open systems. For measurement we still use the mechanical pressure gauges made by Dräger as well as the electronically driven pressure gauges built into such respirators as the Servoventilator (Siemens).

Transcutaneous Blood Gas Measurement and Pulse Oxymetry

Noninvasive measurement of pO_2 and pCO_2 using transcutaneous methods is appropriate not only in infants but also in adults. For such measurements it is necessary that the skin between the blood-carrying capillaries and the membrane of the measuring probe be as thin as possible, and that warming of the electrode to 44°C result in an optimal increase in cutaneous circulation. Decreases in cutaneous circulation due either to blood pressure changes or to the effects of anesthetic gases or vapors in the breathed gases disturb measurements by this technique. Application and adjustment of the measurement probe requires ample time.

For monitoring during anesthesia, the inner side of the eyelid is, at the moment, a sampling point of choice. Such measurements give a good indication of pO_2 and O_2 supply to the eyelid via the internal carotid and ophthalmic arteries, that is, correlating with the O_2 supply to the brain.

Pulse oxymetry (Fig. 18) can also be used to measure transcutaneously O_2 saturation of the blood circulating in the finger tip or the ear lobe. The transcutaneous O_2 and CO_2 measures from the eyelid, together with the other parameters mentioned and combined with the O_2 uptake from the closed-circuit system, can be used continuously to monitor the O_2 supply to the whole body and to the internal organs (liver, kidneys, heart, skeletal muscle, and particularly the brain) without the necessity of invasive measures anywhere on the body.

Naturally, refinements in pulse oxymetry, as well as in transcutaneous O_2 measurements, can and must be made, and these supplemented by the CO_2 measurements which are also possible with such techniques. These techniques could also be exploited to produce pulse-pressure curves, that is, not merely pulse rate but pulse pressure could be directly indicated, and blood pressure values could be followed heartbeat by heartbeat throughout the circulation into the central venous pool. O_2, CO_2, and the pulse-pressure curve together give important information for the evaluation of tissue perfusion, heart work and cardiac output, pulmonary function, and the overall condition of the body.

Given such improvements, which are both possible and necessary, these monitoring possibilities could be realized in conjunction with the closed-circuit system. We have called for such improvements for over 15 years, but developments have taken place in only very small steps.

Cardiac Output. The possibility of measuring O_2 consumption with the closed-circuit system also allows calculation of the cardiac output (CO), i.e., volume of blood pumped per heartbeat or minute. According to Fick's principle: O_2 consumption (ml/min) $\times$ 100 = CO (ml/min) $\times$ $avDO_2$ (vol.%).

The term $avDO_2$ refers to the O_2 difference between the maximal arterial oxygenation of the blood and the maximal deoxygenation at the venous end of the circulation.

As we have seen, O_2 uptake by the patient is measurable in two ways in the closed-circuit system: firstly, as the fresh gas inflow from the dosing apparatus (at the moment, still via rotameter) and, secondly, as the product of the tidal

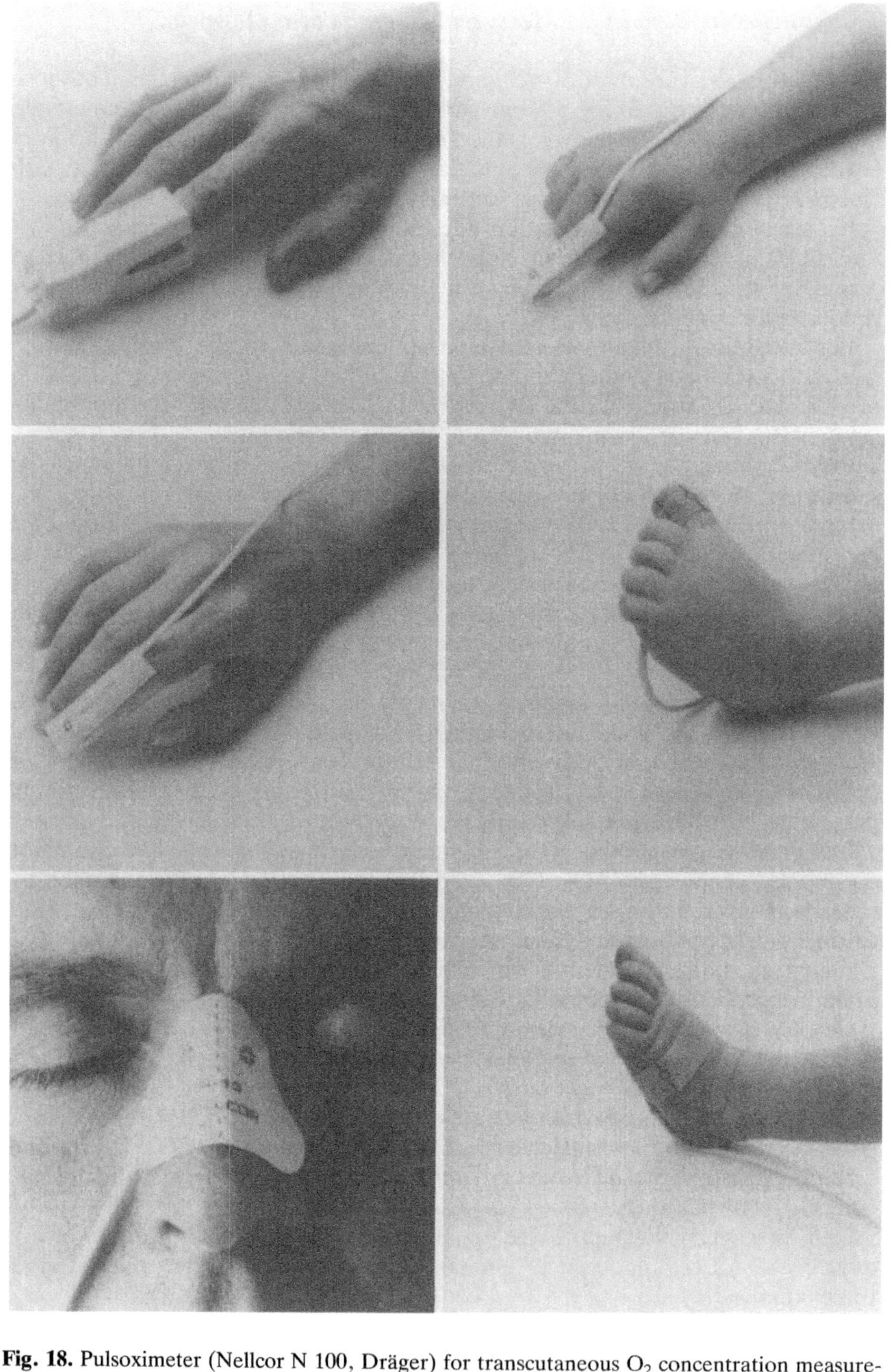

Fig. 18. Pulsoximeter (Nellcor N 100, Dräger) for transcutaneous O_2 concentration measurements of the blood from pulse wave to pulse wave

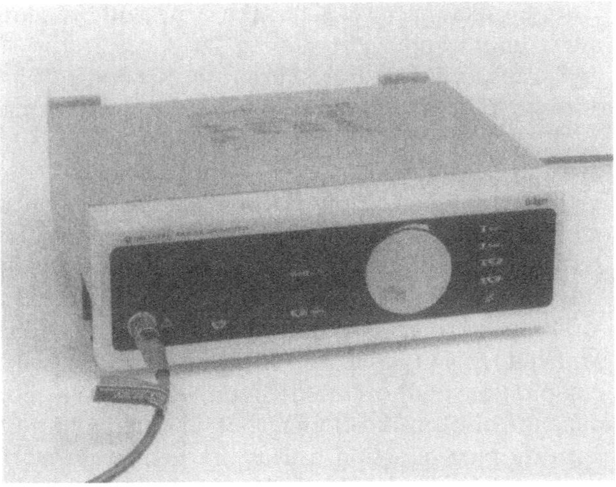

Fig. 18. Continued

volume and the difference in the inspiratory and expiratory O_2 concentration. The left side of the equation shown above, reffering to O_2 consumption, is thus a well-measured parameter. Accordingly, cardiac output can be determined from knowledge of $avDO_2$ or vice versa. Cardiac output may be determined from the uptake and release by the body of inert gases such as N_2O, He, or Xe. It can even be approximated from halothane uptake and release.

The $avDO_2$. The measurement of $avDO_2$ is of interest regarding gas transport in the lungs. In the inspiratory phase, a measured amount of O_2 (as described above) from the atmosphere outside the lungs passes the alveolar membranes (the boundary between the lungs and the blood) into that amount of blood (given by the cardiac output) residing in the capillaries behind this boundary, and a corresponding amount of CO_2 passes out of the blood in the opposite direction across this boundary and into the external atmosphere. The $avDO_2$ is an important descriptive parameter of this process.

For example, if O_2 uptake is 240 ml/min and $avDO_2$ is 4 vol.%, cardiac output is $(240 \times 100) / 4 = 6000$ ml/min. At a pulse rate of 60 heartbeats/min there would be 100 ml blood per heartbeat transported as cardiac output. Conversely, $avDO_2$ may be calculated if O_2 uptake and cardiac output (from the appearance and dilution curves for N_2O, He, or Xe) are known.

The clinical and mathematical correlation of values for cardiac output and $avDO_2$, as well as exact pulse-pressure curves obtained from oxymetry, can be demonstrated vividly. This would be a useful further development of the use of O_2 pulse (O_2 uptake/heartbeat), pulse oxymetry, and pulse-pressure curves, not only in the analysis of the functions of heart and lungs, circulation, and respiration but also for the evaluation of the balance between O_2 uptake and CO_2 release, the evaluation of alveolar ventilation, dead speace and cardiac output, and further the O_2 and CO_2 saturation of the blood, the $avDO_2$, O_2 and CO_2 concentrations of breathing gases, and finally the inspiratory and expiratory concentrations and amounts of the volatile anesthetics.

As an example we may take a patient weighing 70 kg and consuming 240 ml O_2/min, which is taken up at a respiratory frequency of 12 breaths/min and transported with a heart rate of 60/min. The normal value for $avDO_2$ is 4 vol.%. The O_2 pulse is 4 ml O_2, and the cardiac output/beat is 100 ml. With purely carbohydrate metabolism, this 240 ml O_2/ml is metabolized to 240 ml CO_2/min. This 240 ml/min each of O_2 and CO_2 is transported by a respiratory minute volume of 6000 ml and a heart minute volume of 6000 ml. That means that at rest and with a normal $avDO_2$, O_2 uptake and CO_2 release takes place with the same respiratory and heart minute volumes in the lungs.

The performance of the heart, lungs, and forces of diffusion are reflected in $avDO_2$, CO_2 release, O_2 pulse, and cardiac output. This means that exploitation of the closed-circuit system as a noninvasive monitoring system depends also on developing our understanding of the implicit possibilities.

If atropine injection causes the heart rate to rise from 60/min to 120 min/- while $avDO_2$ and O_2 uptake remain constant at 4 vol.% and 240 ml/min, respectively, the O_2 pulse must drop from 4 to 2 ml/heartbeat and the cardiac output/beat from 100 to 50 ml/heartbeat. If, however, O_2 uptake decreases over the course of the anesthesia to a level two-thirds of that needed for basal metabolism, as the result of reductions in body temperature, metabolic rate of inner organs, and muscle tone (muscle relaxation), this is not a sign of a (pathological) reduction of cardiac output but rather an indication of a reduced demand on the heart, as the O_2 requirement of the body has decreased. This is a valuable piece of information which is not available with conventional anesthesia methods. Cardiac arrest during the course of anesthesia is often the result of an excessive toxicological demand rather than a pathological loss of pumping capacity. The facts necessary to recognize such a situation are supplied, among others, by the monitoring system, the closed-circuit system, with its quantitative data concerning the amounts of anesthetic taken up the patient.

With the parameters available from ECG, pulse-pressure curves, CO_2 production, O_2 and CO_2 concentrations (saturations) in the blood, and alveolar and dead-space ventilation, one may in any case differentiate among infarct, tachycardia, arrhythmies, hypovolemia, disturbances of acid, base, water, and electrolyte balances, cardiac insufficiency, and reduced O_2 requirement of the anesthetized organism as the cause of a particular decrease in cardiac output.

Knowledge of O_2 consumption is a piece of information with far-reaching implications, but one which is lost when nonclosed anesthesia systems are used. The new parameter, O_2 consumption, should not only be accorded an importance on a level with the other parameters which we have used up to now (blood pressure ECG, pulse, pO_2, pCO_2, tidal volume, respiratory frequency, respiratory minute volume and pressures, concentrations and amounts of O_2, CO_2, and anesthetic gases) but seen rather as a parameter which could substantially change our clinical and pharmacologic procedures – from the application of premedication, such as atropine and neuroleptics, to the postoperative management of pain with analgetics.

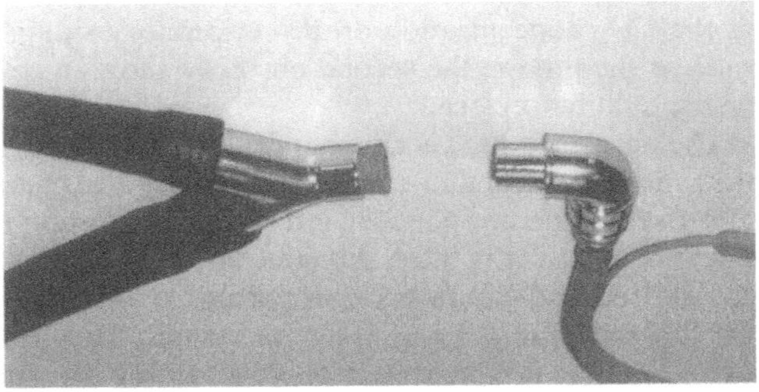

Fig. 19. Closure of the system at the end of the Y piece with a rubber stopper, to allow absorber lime and rubber parts of the system to become saturated with volatile anesthetic. The system is so closed roughly 15 min before anesthesia is to begin and also serves as a check that the system is gas tight

CO_2 Absorption Mass

Exhausted CO_2 absorption mass may be recognized by color change, appearance of CO_2 on the inspiratory side of the circuit, or by the lack of reaction warmth. Its use is indispensable for the closed-circuit system. Soda lime (white) and barium lime (pink) are most commonly used. We generally use soda lime, which is composed principally of calcium hydroxide with small amounts of hydroxides of sodium and potassium. The absorption of CO_2 by $Ca\ (OH)_2$ yields $CaCO_3 + H_2O$ + heat. The produced heat and water vapor significantly affect the temperature and humidity of gases in the closed-circuit system. CO_2 absorption by soda lime is very effective at a humidity of about 17%.

One gram of soda lime is contained in a volume having a surface area of $13\,cm^2$, but the total surface area of the crystals in this 1 g amounts to about $560\,cm^2$. The uptake capacity of the lime is dependent upon the surface area of the crystals. The soda lime in the Absorbertopf (Dräger) has a theoretical absorption capacity of 240 l CO_2. When the soda lime becomes exhausted, CO_2 can pass through to the inspiratory branch. To prevent this we use two absorbers connected in series. In this manner the absorber mass in one unit may be virtually exhausted without CO_2 passing through into the inspiratory branch.

The resistance of a second absorber leads to a pressure fall of about 0.5 mbar at a flow of 60 l/min and may therefore be neglected.

Ethyl violet, which is added to the soda lime as a pH indicator, changes from white to violet at a pH of 10.3.

After longer storage, the indicator can give off an ammonia-like smell. This smell may be quickly dispersed without loss of absorption capacity by passing fresh air through the filter. The color change of the indicator is reversible, as long as the soda lime is not totally exhausted (saturated with CO_2). The absorbing mass is not exhausted when the indicator first changes color. It regenerates to some extent after a time. It may be used further as the first element in series with a second absorber containing unexhausted lime, as long as the CO_2 concentration on the inspiratory side is constantly monitored. As soon as the lime in the second absorber shows a color change, the mass in the first should be replaced.

The absorbing lime is not chemically affected by the various volatile anesthetics, with the exception of trichlorethylene and chloroform. It should be noted that the volatile anesthetics bind to the absorption mass, as does water vapor. For this reason it is often difficult, at the beginning of anesthesia, to maintain the desired end-expiratory concentrations of volatile anesthetics with the low gas flows used in the closed-circuit system. Thus, particularly at the beginning of anesthesia, it is necessary to work with high concentration settings for the anesthetic vapors. We have found it extremely helpful to fill the system with the anesthetic gas mixture (O_2, N_2O, and volatile anesthetic) and close it about 10 min before anesthesia is to begin (Fig. 19). The system is most easily closed by placing a rubber stopper in the end of the Y connector.

During this period the volatile anesthetic can permeate the rubber parts and the absorbing mass. When anesthesia is to begin, it is necessary, in addition, to flood the system with large amounts of fresh gases for a short period (1–3 min) with the pressure-relief valve more or less open.

One must also remember that the rapid rise in water vapor pressure in the closed-circuit system, as described above, releases a part of the volatile anesthetic that was bound by the absorption mass. Since we continuously monitor the end-expiratory and inspiratory concentrations of volatile anesthetic, this rise is noticed immediately, and an overdose can be avoided by reduction of the vapor inflow. Large changes in dosing are not necessary, as the addition of volatile anesthetic can be as finely and exactly controlled as that of the anesthetic gases.

Dräger specifies a shelf-life for their absorption lime of up to 5 years at storage temperatures between −20° and +50°C.

Temperature of Anesthetic Gases

We measure gas temperatures in the inspiratory and expiratory branches using instruments made by the Yellow Springs Instruments Co. (Ohio, USA) and Rotronic (Frankfurt/Germany). The gas temperature sinks from body temperature of the patient (36°C) down to room temperature (22°C) in its course from the expiratory to the inspiratory branch. The better the entirety of the closed-circuit system is insulated against loss of heat, the less this drop in temperature is. With exact temperature measurements it is possible to detect not only a loss of absorption capability of the lime (heat of reaction) but also whether the metabolic rate is changing slowly or quickly (e.g., in hyperthyroidism and hyperthermia), and whether the patient is losing heat to the anesthetic gases. Measurement of the temperature of the anesthetic gases is as important as the determination of gas pressures in the system for the calculation of gas volumes and concentrations down to the molecular level. It is also true in this case that the more accurate the measurements are, the more accurate the calculatuions will be, and the more finely the anesthesia may be dosed.

Neuroleptic Anesthesia

During induction the patient is respirated via a mask with a fresh gas flow of approximately 6 l O_2/min with more or less open pressure-relief valve and tightly fitting mask. As a rule, this phase lasts some 3 min, and the proportion of N_2 in the lungs can be quickly reduced from 79 vol.% down to a few volume percent, being replaced with pure O_2. Next, roughly 75 vol.% of the pure O_2 is replaced with N_2O, the analgesic properties of which allow a reduction in the amount of fentanyl necessary. If the N_2 concentration in the lung gases is not yet sufficiently reduced at the time of intubation, one can either (a) add pure O_2 in excess to the system, with the pressure-relief valve more or less open, until the O_2 concentration in the lungs and closed-circuit system approaches 100 vol.%, or (b) apply a high (6 l/min) fresh gas flow of (1.5 l/min) O_2 and (4.5 l/min) N_2O, that is, 75 vol.% N_2O and 25 vol.% O_2, with the pressure-relief valve more or less open, until the desired volume percent values for N_2O, O_2, etc., have been reached.

As we know, the analgesic effects of N_2O increase with the concentration of N_2O in the gas mixture. A value between 70 and 75 vol.% is optimal. The N_2O concentration may be increased more economically, if more slowly, from 0 vol.% to 75 vol.% by cutting off the fresh gas inflow of O_2 and giving the necessary fresh gas as pure N_2O. This requires a time span which is dependent, of course, on the O_2 consumption of the patient and on the total volume of the system.

By way of an example, we can return to the above male patient, 170 cm in height, weighing 70 kg, and having a total capacity of 5000 ml, and consider a closed-circuit system, also having a gas volume of 5000 ml. The total volume of the gas circulating between the closed-circuit system and the patient is thus 10000 ml. If the concentration of O_2 in the circulating gas is 100 vol.%, the patient consumes 245 ml O_2/min at rest, and the desired concentrations are 75 vol.% N_2O and 25 vol.% O_2, then the time required in order for the patient to use up 75 vol.% O_2 from the gas volume of 10 000 ml may be calculated as follows: 75 vol.% 10 000 ml O_2 = 7500 ml O_2; 7500 ml/245 ml/min = 30 min, 37 s.

During this 30 min and 37 s, 7500 ml O_2 is metabolized, and the amount of O_2 is reduced from 100 vol.% to 25 vol.%, 7500 ml O_2 being absorbed as CO_2 by the soda lime. This means, of course, that 7500 ml will be breathed up by the lungs and closed-circuit system if N_2O is not added to the system, and the O_2 concentration of the remaining gas is still 100 vol.%. This, however, was not our intention. We wanted to attain gas concentrations of 75 vol.% N_2O and 25 vol.% O_2. To this end, the 245 ml O_2/min breathed up must be replaced with 245 ml N_2O/min added as fresh gas. This figure does not allow for the amount of N_2O lost to the atmosphere via skin, intestines, and breathing tubes, which must also be replaced. Additional N_2O inflow is thus needed. As we saw above, this amount may be calculated as 1.5 ml/kg adjusted body weight. In the present case this amounts to 70 × 1.5 = 105 ml/min. Thus, we must add 245 ml/min + 105 ml/min = 350 ml/min N_2O during the 30 min and 37 s, in order to achieve a gas mixture of 25 vol.% O_2 and 75 vol.% N_2O from the starting point of 100 vol.% O_2 in the 10000 ml circulating gas.

Such calculations are, of course, to be continually checked against actual measurements. The appropriate control of anesthesia is, in the final analysis, dependent upon measurement data, not model calculations. In the future, however, computer-assisted evaluation of such data will become increasingly important and helpful for clinical anesthesia procedures.

This example shows that the necessary analgesic concentration of N_2O is attained very slowly with closed-system substitution methods. For this reason, the use of high fresh gas flows (system temporarily partially open) of, for example, 75 vol.% N_2O and 25 vol.% O_2 at rates of 4500 ml N_2O/min and 1500 ml O_2/min are more or less unavoidable in clincial work in order to wash the pure O_2 out of the system and lungs.

Since we consider the presence of N_2 in the breathed gases to be of importance in protecting the alveolar surfactants, we do not attempt particularly to wash out all N_2 in neuroleptic anesthesia either. We consider a N_2 concentration of 5–7 vol.% to be advantageous. This means that our starting point is not 100 vol.% O_2 in the total gas volume of 10000 ml, but rather a correspondingly lower O_2 volume percent value.

We continue the neuroleptic anesthesia in the normal fashion with intermittent intravenous injection of fentanyl. Occasionally we supplement with additional injections of small amounts of thiobarbiturates (e.g., 125 mg Trapanal). Technically, the neuroleptic anesthesia with the closed-circuit system is carried out exactly as is done with nonclosed systems.

Termination of Anesthesia

If anesthesia must be quickly terminated, a high O_2 fresh gas flow (e.g., 6000 ml/min) may be used with a more or less fully open pop-off valve. In this way the anesthetic gases are quickly washed out of the system and substituted by pure O_2. Volatile anesthetics such as halothane, enflurane, and isoflurane can be removed particularly quickly when an absorber unit filled with charcoal is introduced into the circulating gas stream.

If the termination of anesthesia is to be more gradual, the system is left completely closed and the inflow of volatile anesthetic vapors and N_2O is switched off. The supply of fresh O_2 is continued to the system in the amounts needed by the patient. So that the total amount of circulating gas does not decrease because of release of N_2O from the patient's body to the atmosphere, an additional fresh O_2 flow is given, equal in magnitude to the previous N_2O flow. That is, the total gas flow during termination of anesthesia must remain approximately the same as it was during anesthesia. The only difference is that fresh gas flow during termination consists of pure O_2. Using this technique, the termination phase lasts, on average, between 15 and 30 min, and the duration may be shortened or prolonged at will. At any point, the period of anesthesia can be shortened by increasing the fresh O_2 flow and opening the pop-off valve. Within limits, the higher the flow, the more quickly the system will be washed out, and the sooner the anesthesia will terminate. On the other hand, anesthe-

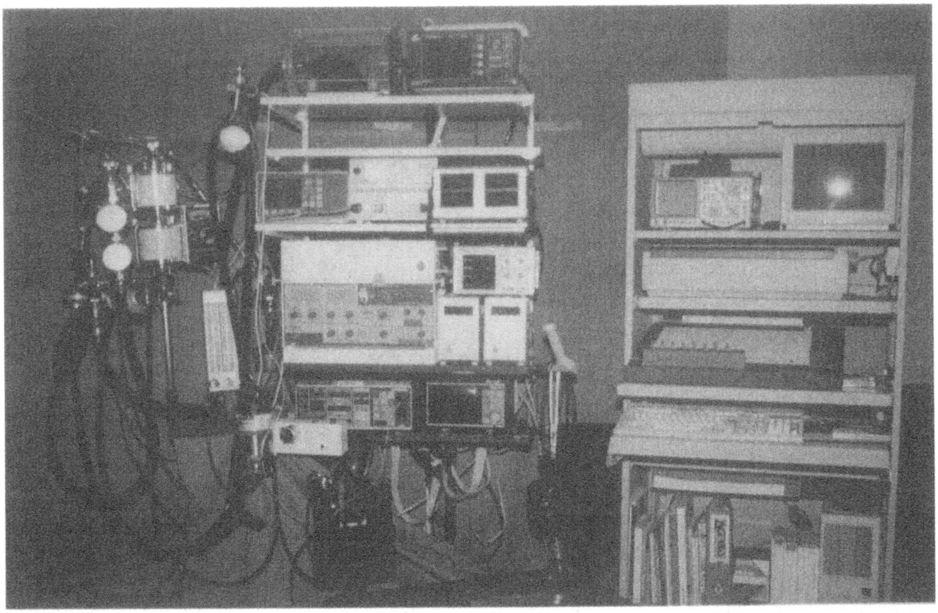

Fig. 20. The closed-circuit system in its entirety with the measuring and processing instruments as we use it

sia may be deepened by reintroducing flows of N_2O and volatile anesthetic and/or giving intravenous anesthetics.

The closed-circuit system allows the same fine control over the termination phase as it provides for the other phases of anesthesia.

Summary

The closed-circuit system is itself a monitoring system which indicates, at every point, the amounts and uptakes of O_2, N_2O, the noble gases, and vapors such as halothane, enflurane, and isoflurane (Fig. 20).

If instruments for measuring the concentrations and volumes of the gases and vapors are built into the closed-circuit system, this monitoring systems becomes a feedback-control system, as the uptake and delivery of the anesthetic gases and vapors on the inspiratory and expiratory sides are mutually interdependent.

The superiority of the closed-circuit system over half-closed and open systems is manifest in every respect, from its monitoring function per se to the fully automatic control and regulation of the anesthesia as well as the standard of safety attained by its use.

Since the delivery, addition, or uptake of O_2, N_2O, halothane, isoflurane, and noble gases may be controlled as exactly as the CO_2 production, not only in their amounts but also in their concentrations, the closed-circuit system lends itself as the perfect, and handy, respiration control system.

New parameters, in terms of units per breath or heartbeat, arise from the temporal correlation of uptake and delivery of O_2, N_2O, halothane, isoflurane, enflurane, and noble gases with hemoglobin, hematocrit, red cell count, age, body temperature, muscle tone, height, body weight, pulse-pressure curves, etc. These parameters represent a further development of the O_2 pulse and make possible more accurate evaluation of the work and capacity of the heart and lungs.

The use of pulse oxymetry, pulse capnography, and the transcutaneous evaluation of tissue pO_2 and pCO_2, in addition to the measurements listed above of the gases circulating between the lungs and the closed-circuit system, are contributing to the development of an important further differentiation of feedback control.

Furthermore, it is surely to be expected that future technical advances in the area of transcutaneous noninvasive measurements will be made, such that the determination of N_2O, N_2, CO_2, noble gases, and anesthetics will be possible in a fashion analogous to that of pulse oxymetry and capnography. Perfection will, however, be reached only when it is possible to obtain all the necessary data in a noninvasive manner, both from the lungs and from the blood, and when computer-assisted processing of this data allows the determination of the new parameters down to the molecular level for the individual organs in terms of milligrams.

This is the road to a "molecular anesthesia," and the first step on this road is the use of the closed-circuit system. Developments in this area will have an impact on the whole of intensive care, including the treatment of shock, burns, poisoning, resuscitation and metabolic diseases, as well as on the development of space medicine. This work will also set the pace for a terrestrial and extraterrestrial molecular medicine.

For anesthesia, the closed-circuit system is the system which allows a complete recycling of the expired anesthetic gases and vapors, and which also makes possible the introduction of expensive noble gases such as Xe for use as analgesics.

References

Droh R (1976) Das geschlossene Narkose-Kreissystem. Anaesth prax 11:27–32

Droh R (1976) Clinical significance of recent research results. (VIth World Congress of Anaesthesiology, Mexico, April 24–30, Programa official: p 84, 369)

Droh R, Rothmann G (1977) Das geschlossene Kreissystem. Eine diskreditierte Konzeption der Vergangenheit, durch Innovationen die Methode der Wahl. Anaesthesist 26:461–466

Droh R (1981) The closed circuit system and its advantages: a review of the last nine years. (International Symposion on Closed Circuit Anaesthesia, Birmingham/USA, 10.–12. April)

Droh R (1982) Die Inhalationsnarkose im geschlossenen System. (Internationales Symposion: Geschlossenes System für Inhalationsnarkosen, Düssedorf, 7. und 8. Mai)

Droh R, Spintge R, Erdmann W (1982) Die Überlegenheit des geschlossenen Kreissystems. Ein Zehnjahresbericht über 50000 Fälle (Deutscher Anaesthesie-Congress 1982)

Droh R (1982) Closed circuit anaesthesia and it's significance. (VIth European Congress of Anaesthesiology, London, September, 8–15)

Droh R, Erdmann W, Spintge R (1985) Anaesthesia – innovations in management. Springer, Berlin Heidelberg New York Tokyo

Droh R, Spintge R (1986) Closed circuit system and other innovations in anaesthesia. Springer, Berlin Heidelberg New York Tokyo

Gauer OH, Kramer K, Jung R (1971) Physiologie des Menschen, Bd 2: Energiehaushalt und Temperaturregulation. Urban & Schwarzenberg, München Berlin Wien

Lowe HJ, Ernst EA (1981) The quantitative practice of anaesthesia, use of closed circuit, Williams & Wilkins, Baltimore/USA

Lawin P (1981) Praxis der Intensivbehandlung. Thieme, Stuttgart New York

Lachmann B (1986) Zur Xenonanästhesie und Konstruktion des „Bag-in-box-Systems“. (Persönliche Mitteilung)

Determination of Anesthetics in Aqueous Solution by Infrared ATR Spectrometry

R. WYZGOL, P. HEINRICH, B. SCHRADER, and D. W. LÜBBERS

Introduction

One of the priority tasks in anesthesiology is the continuous monitoring of the uptake and elimination of anesthetic during and after anesthesia. The depth of anesthesia depends on the concentration of anesthetics in the brain. Since this is in equilibrium with the concentration in the blood, the determination of the content of anesthetics in the blood is a valuable method for obtaining information about the depth of anesthesia. This can be done by means of infrared spectrometry. The very intensive absorption bands shown by the commonly used volatile anesthetics halothane, enflurane, isoflurane, and gasous nitrous oxide (N_2O) allow the determination of these compounds. Respiratory gas analysis under clinical conditions by infrared spectrometry is well known, using the ordinary transmission technique. In our view, the parameters of blood analysis correlate more significantly with the depth of anesthesia.

The difficulty in the case of blood is its high content of water and other strongly absorbing substances that disturb the transmission spectrum. A suitable possibility to use infrared spectrometry to investigate strongly absorbing samples is so-called ATR (attenuated total reflection) spectrometry, based on the optical phenomenon of reflection. It can be observed when light strikes the interface between two media of different indices of refraction. Total reflection only exists if light propagating in a medium with a higher index of refraction n_1 strikes the interface to a medium with a lower index of refraction n_2 (Fig. 1). In this case the angle of incidence Θ must be larger than the critical angle, which is given by $\sin \Theta_c = n_2/n_1$.

According to Maxwell's theory of electromagnetic waves [1], radiation penetrates into the "rarer" medium during total reflection (Fig. 2). As a result of superposition of incoming and reflected beams, there is a sinusoidal standing wave perpendicular to the interface in the denser medium; an "evanescent" wave results in the rarer medium. Its electric field vector $\vec{E}$ decreases exponentially with increasing distance z from the interface.

$$\vec{E} = \vec{E}_0 \exp\left(-z/d_p\right)$$

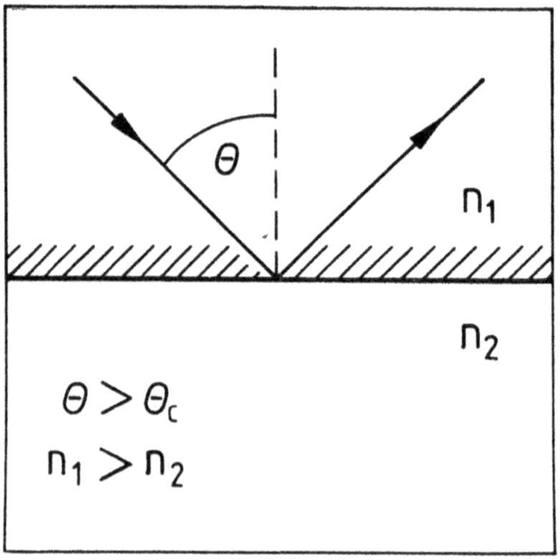

Fig. 1. Total internal reflection. n_1, Index of refraction of the denser medium; n_2, index of refraction of the "rarer" medium; Θ, angle of incidence; Θ_c, critical angle

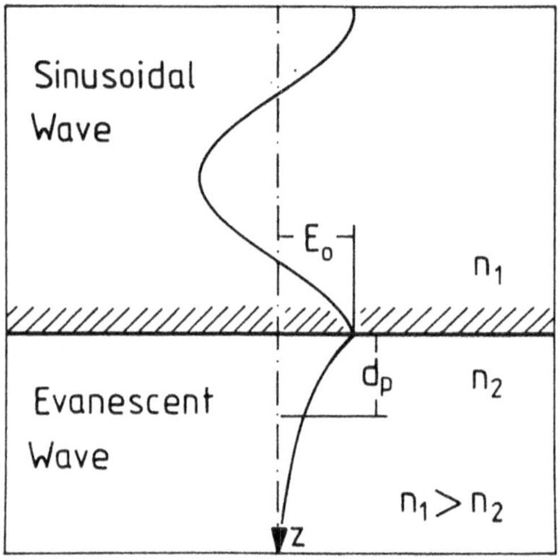

Fig. 2. Standing wave near a totally reflecting interface. E_o, Electrical field vector on the interface; d_p, penetration depth

Here $\vec{E}_o$ is the amplitude of the electric field vector on the surface, z the distance of the evanescent wave from the interface, and d_p the so-called penetration depth given by the distance where $\vec{E}$ is 1/e of $\vec{E}_o$:

$$d_p = \frac{\lambda}{2\pi n_1 \left(\sin^2\Theta - n_{21}^2\right)^{1/2}} \quad ; \quad n_{21} = n_2/n_1$$

d_p depends on the wavelength λ of the light, the angle of incidence Θ and the indices of refraction. The order of d_p is in the range of 0.1–1 λ.

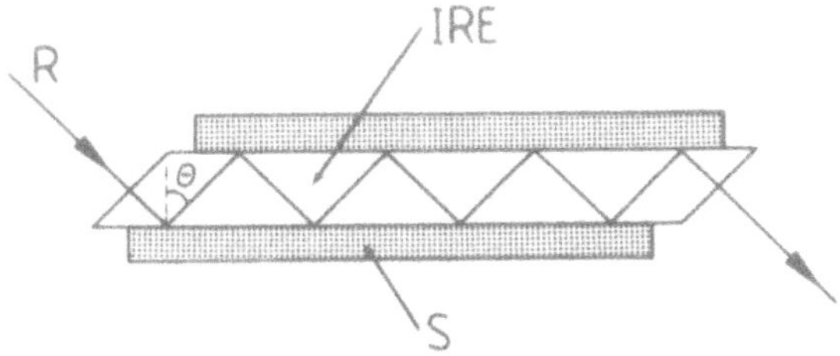

Fig. 3. Commonly used internal reflection element *(IRE)*. *R*, Radiation; *S*, sample

In the case of an absorbing "rarer" medium (sample) the evanescent wave interacts with it, and the reflected radiation loses a portion of its energy. This phenomenon is called attenuated total reflection. For the experimental application of the ATR technique the infrared radiation enters an internal reflection element (IRE) of an appropriate infrared transparent material, where it is totally reflected at the inner surface. The IRE may be shaped like a cylinder, trapezoid, or parallelopiped. The latter is shown in Fig. 3. The sample is placed on the polished lateral surfaces of the IRE. Solids are mechanically pressed onto the surfaces; liquids wet the surfaces in special cuvettes. Most organic and biological samples have an index of refraction in the range of 1.3–1.7. For this reason the index of refraction of the IRE should be above 2. Suitable materials include zinc selenide (ZnSe), zinc sulfide (ZnS), germanium (Ge), and silicon (Si). The ATR effect is amplified when multiple reflections are used. The number of reflections N inside the IRE depends on the length l and the thickness t of the element and is given by: $N = l/t \cot \Theta$. At each reflection the evanescent wave penetrates into the sample medium, and the absorption spectrum of the sample – equivalent to its transmission spectrum – can thus be observed.

A further difficulty for the determination of anesthetics in the blood is the very low concentration to be determined. The interesting range of concentration in anesthesiology lies between 50 and 200 mg/l. This is not possible with commonly used ATR spectrometry, and it is because of this that we have developed a modification of the ATR technique. It is well known that volatile and gasous anesthetics have a much higher solubility in polymers such as silicon rubber, polyvinyl chloride or butyl rubber than they have in water. Therefore, we coat the IRE with a polymer membrane which enriches the anesthetics with respect to its concentration in the solution [2]. The thickness of the membrane is larger than the depth of penetration dp, as is shown schematically in Fig. 4. Thus, the evanescent wave interacts only with the membrane itself and the enriched anesthetic therein. Since the membrane is hydrophobic, it promotes the enrichment of the anesthetic and detains the other components of the blood. Only hydrophobic components, e.g., anesthetics or gases, diffuse into the membrane until the equilibrium of distribution between the concentrations in the solution and the membrane is reached, which is theoretically about 1:100. The high content of anesthetic in the membrane lowers the limit of detection and gains the sensitivity drastically. It should be noted that the membrane material must be transparent in the interesting regions where absorption bands of the anesthetic occur.

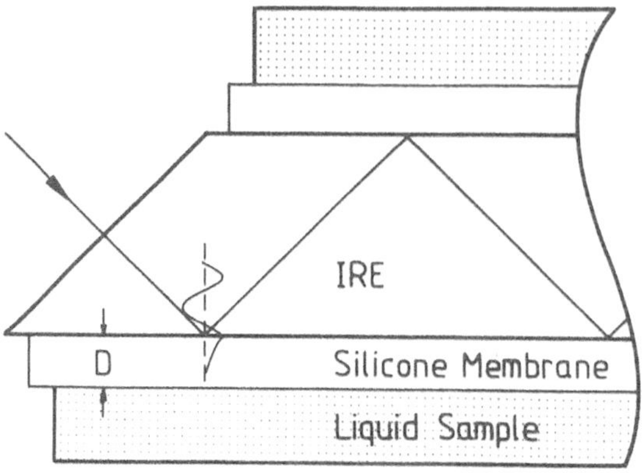

Fig. 4. Schematic representation of coated reflection element.
D, Thickness of the membrane;
D, $> d_p$

Results and Discussion

The analytical method described above has been successfully used for determination of the volatile anesthetics halothane, isoflurane, and enflurane in aqueous solution in the anesthesiologically interesting range of concentration. In the following, the efficiency of this method is shown on halothane representatively [3].

The ATR spectra were recorded with a Fourier-Transform infrared spectrometer (Nicolet 7199 FTIR) co-adding 64 interferograms at 2 cm^{-1} resolution. An ATR accessory of Barnes Analytical (Model 300) and parallelogram-shaped internal reflection elements of germanium and zinc selenide with an angle of incidence of 30° and 45° were used. The number of reflections was 28 and 16, respectively. Halothane was obtained from ICI-Pharma (Plankstadt, FRG). Aqueous solutions of halothane passed continuously the coated reflection element while recording the spectra. The membrane was generated of two component polydimethylsiloxane (Silgel 603 from Wacker-Chemie, Munich, FRG) directly on the reflection element.

There are two important reasons to use this membrane material. First, polydimethylsiloxane, a silicon rubber, has a high receptivity for halothane, recognizable in the high Ostwald's solubility coefficient of 98. Second, as shown in Fig. 5, there is a "window" in the infrared spectrum of polydimethylsiloxane between 1413 cm^{-1} and 1270 cm^{-1}, allowing the evaluation of the absorption band of halothane at 1310 cm^{-1} (7.6 μm).

Figure 6 shows plots of the halothane peak absorbance at 1310 cm^{-1} versus the concentration in water. The measurements were carried out with an uncoated and with a coated germanium element. In the former, the results are not satisfying because the limit of detection is about 1500 mg/l and does not reach the concentration range of interest. Only concentrations above 1500 mg/l can be identified with a statistical accuracy of 99.8%. The slope for the uncoated is 55 times larger than that for the coated element, indicating that the enrichment factor of halothane in the silicon membrane is also approximately 55. Hence, ATR spectra of a very high signal-to-noise ratio can be obtained, and the limit of detection is lowered to about 40 mg/l.

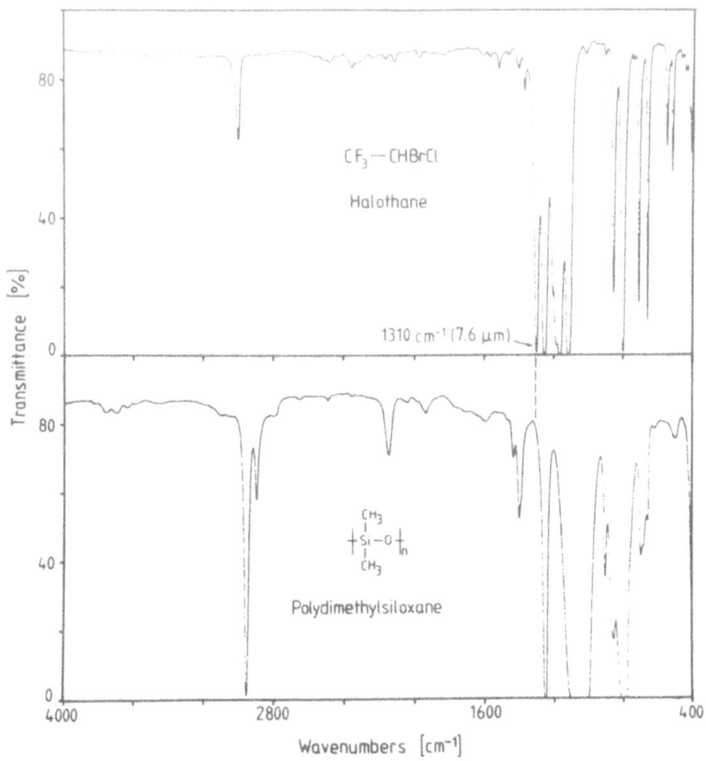

Fig. 5. Infrared spectra of halothane ond polydimethylsiloxane (membrane material)

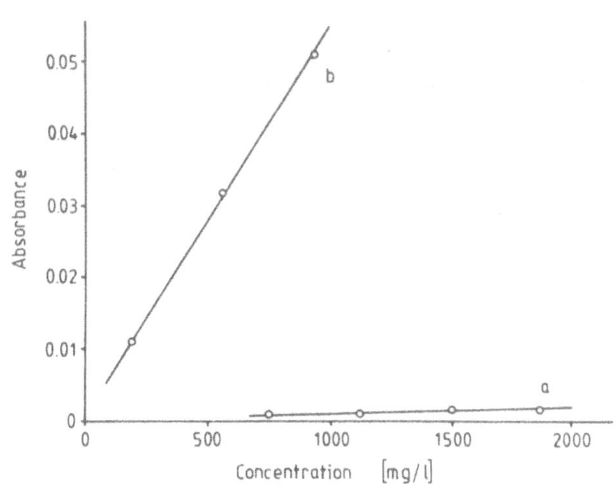

Fig. 6. Peak absorbance of the 1310 cm^{-1} band of halothane against the concentration in aqueous solution, measured with an uncoated *(a)* and coated *(b)* germanium IRE

To improve this method a reflection element made of zinc selenide can be used; this has a higher transmissivity for infrared radiation and a lower index of refraction than germanium ($n_{ZnSe} = 2.4$; $n_{Ge} = 4.0$). Its higher transmissivity

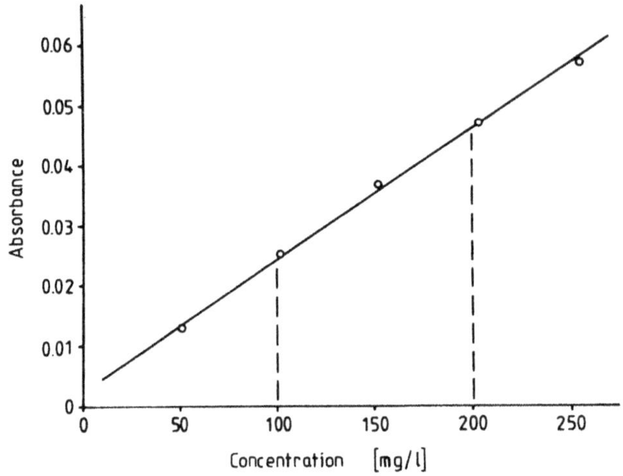

Fig. 7. Peak absorbance of the 1310 cm^{-1} band of halothane against the concentration in aqueous solution, measured with a coated ZnSe IRE. (Concentration in blood during anesthesia, 100–200 mg/l)

increased the signal-to-noise ratio, and its lower index of refraction increases the sensitivity causing a larger penetration depth d_p.

The peak absorbance of the halothane band detectable with a coated ZnSe element is plotted in Fig. 7 against the concentration in the anesthesiological range of concentration, which lies between 100 and 200 mg/l.

The signal-to-noise ratio permits a limit of detection of less than 10 mg/l. The linear dependence between absorbance and concentration indicates that the Lambert-Beer law is valid for ATR spectrometry in this range of concentration. The ATR spectra of halothane in aqueous solution shown in Fig. 8 also give epxression to this fact.

A continuously or semicontinuously working analytical method is characterized by its response time. In our case the response time is given by the time which the anesthetic needs to diffuse into the silicon membrane. To determine this time, constant τ rapid-scan spectra were recorded for different membrane thicknesses. τ is defined as the time in which absorbance decreases to $1/e$ of the initial value or increases to $1-1/e$ of the final value, respectively.

For membrane thicknesses of 60 ± 10 μm a time constant of about 2 min results, whereas for 25 ± 5 μm it is only 1 min ± 10 s. As mentioned above, the penetration depth d_p is approximately in the range of the wavelength λ, which is 3–10 μm in the interesting region of infrared radiation. Hence, an optimized membrane thickness would be in the same range. Such membranes would have a time constant between 5 and 30 s.

Conclusion and Future Developments

ATR infrared spectrometry using coated internal reflection elements has been shown to be a suitable technique for the determination of halothane in aqueous solution in the range of concentration between 10 and 250 mg/l. The same procedure is applicable to enflurane, isoflurane, nitrous oxide, and even to

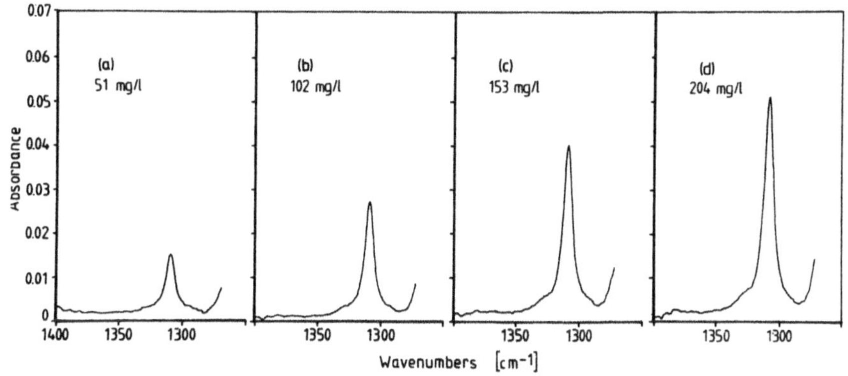

Fig. 8. ATR spectra of halothane in aqueous solution, measured with a coated ZnSe IRE

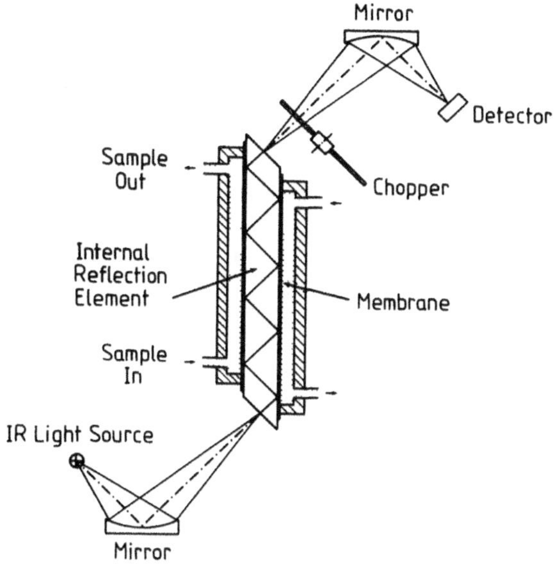

Fig. 9. Schematic outline of a routine filter spectrometer for continuous measurements

physically dissolved carbon dioxide, as we have recently shown [4]. At this time we are confident that this method can be used to determine volatile and gasous anesthetics in the blood. The FTIR spectrometer used in this study, however, is not a routine apparatus, but a simple routine filter spectrometer for continuous measurements under physiological conditions will be constructed and tested.

As is shown schematically in Fig. 9, only four constructional units are necessary: an infrared light source, a coated reflection element, a rotating beam chopper, and a detector. The chopper contains two narrow bandpass filters, a measuring filter and a reference filter. The measuring filter is transparent only for the wavelength of an appropriate absorption band; the reference filter selects a wavelength in which no absorption occurs. The ratio of both signals is proportional to the absorbance of the band. A chopper containing more than two filters can be applied for simultanuous measurements, e.g., of halothane and nitrous oxide.

Further developments are in progress to optimize the ATR flow cell and to develop new membrane materials.

References

1. Harrick NJ (1967) Internal reflection spectroscopy. Interscience, New York
2. Opitz N, Lübbers DW, Schrader B (1983) Anordnung zur optischen Messung von Stoffkonzentrationen. EU-P 144713A2 (06. 12. 83)
3. Wyzgol R (1985) Quantitative Bestimmung der Inhalationsanästhetika Halothan und Enfluran mit Hilfe der ATR-ITR Spektrometrie. Thesis, University of Essen
4. Heinrich P (1983) CO_2-Bestimmung von wäßrigen Medien mit Hilfe der ATR-IR-Spektroskopie. Thesis, University of Essen

Experiences with a New Wavelength Scanning Oximeter

D. W. Lübbers, U. Merschebrock, and H. P. Keller

It is well known that hemoglobin changes its color during oxygen uptake and oxygen release (see [1]). As early as the last century an attempt was made to use the hemoglobin in blood vessels of the skin as an indicator for the degree of oxygenation of blood in order to draw conclusions on characteristics of blood or skin. After the invention of photoelements, ear and finger oximeters for determining the HbO_2 content of blood were developed following investigations of Kramer [2], Matthes [3–5], and Millikan [6], (for reviews see [6–10]).

Figure 1 shows the changes in the hemoglobin absorption spectra during release of oxygen. The two-peaked spectra correspond with hemoglobin totally saturated with O_2 (HbO_2 or oxyhemoglobin). Additionally, the absorption spectra of decreasing O_2 saturation are shown (80%, 60%, 40%, 20%). The deoxygenated blood (0% saruration, deoxyhemoglobin, Hb) has a one-peaked spectrum. It can be seen that the spectra intersect at four wavelengths. These are the so-called isosbestic wavelengths. At these wavelengths extinction does not depend upon the oxygenation but upon concentration of hemoglobin only. Large changes during oxygenation are found within the range of maxima. Therefore, in order to determine the oxygen saturation of hemoglobin, the two-wavelength method was developed, by which the difference of extinctions between isosbestic and measuring wavelengths [3] can be measured. This method allows the elimination of disturbances in case they have identical effects on both wavelengths. Since the two-wavelength method forms only the extinction difference, it fails if the concentration changes. A method independent of concentration was indicated in 1957 [11] which presupposes that the absorption spectra in a larger range of wavelength are known.

The absorption spectra in Fig. 1 can be taken for a two-component system formed by a mixture of its basic spectral components, oxygenated and deoxygenated hemoglobin. Knowing the spectra of the basic components, the O_2 saturation of a mixture can be determined by calculating the concentration of the components (basic spectra) at which the least-squares sum (Gaussian distribution) is a minimum. This statement holds not only for two-component but also for multiple-component systems. In all these measurements it is assumed that following the Lambert-Beer law, which is the basic law of photometry, the negative logarithm of transmission equals the extinction, i.e., all other conditions being the same the extinction follows changes in concentration linearly.

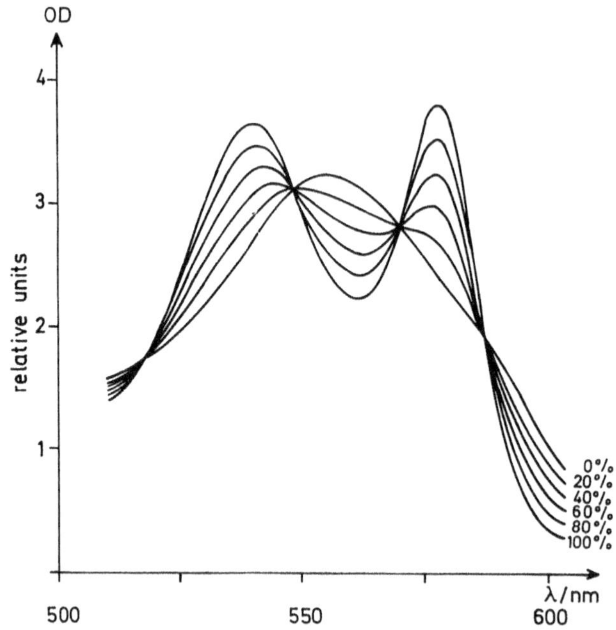

Fig. 1. Spectra of a hemoglobin solution of different degree of O_2 saturations (transmitted light): 100%, 80%, 60%, 40%, 20%, 0%. During deoxygenation the two peaks of 100% O_2-saturated hemoglobin spectrum (577 nm and 542 nm) decrease in amplitude, and the distance between the peaks becomes smaller. At 20% saturation only a broad single maximum is left, which sharpens at 0% saturation (maximum at 555 nm). Isosbestic points are at wavelengths 586 nm, 569 nm, 548.5 nm, 522 nm; extinction in optical density, *OD*

Practical use of the oximeter has shown that the Lambert-Beer law holds only approximately. The individual conditions of measuring may vary so that individual calibration curves must be provided in order to obtain correct results. Unfortunately, this fact inhibits the practical use of the method considerably. While the Lambert-Beer law for transmitted light in the infrared range holds in good approximation, this is not true in reflection measurements [8–10], especially at shorter wavelengths.

We investigated the possibility of overcoming the necessity to work with an individual calibration curve. To find out more about our optical system we analyzed the spectral changes which occur because: (a) the hemoglobin does not distribute in vessels within the tissue as evenly as in a cuvette and (b) in measurements in tissues only reflected light is recorded. We proceeded as follows: local hyperemia (application of Finalgon, a derivative of nicotinic acid) and oxygen respiration, provided that hemoglobin in the blood of skin was saturated with oxygen to practically 100%. If no optical changes occur in reflection oximetry under these conditions, the recorded reflection spectra of hemoglobin in the skin ought to correspond with the absorption spectra of an equivalent HbO_2 solution.

On the left in Fig. 2 the spectrum of the hyperemized fingertip is to be seen with completely oxygenated blood (a). Below on the right (c) the absorption spectrum of a HbO_2 solution of about the same concentration is plotted. Comparing the extinctions at identical wavelengths and plotting them in the coordinate system between the spectra, a distinctly bent curve occurs (b), which flattens and splits towards higher extinctions. With this curve being known, it is possible to transform the measured reflection spectrum into the

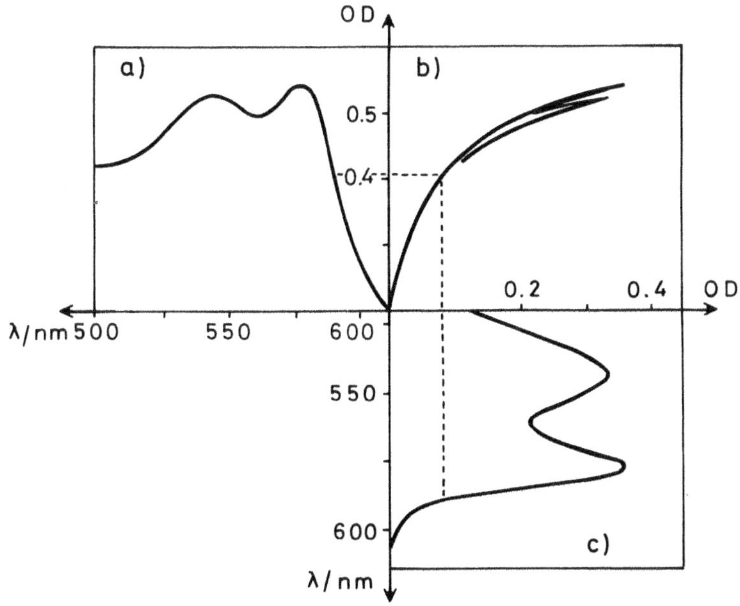

Fig. 2. Transformation function of the blood-perfused skin. The reflection spectrum of the skin with fully oxygenated blood *(a)* is compared wavelength by wavelength *(dotted lines in b)* with the spectrum of an HbO_2 solution. This comparison results in the transformation curve *(b)* which characterizes the optical system to be measured

true absorption curve of hemoglobin. Therefore, we called this curve "transformation function" *(H)*. The strong bend of the transformation curve is caused by the nonhomogeneous distribution of blood in the skin and by scattering [12, 13]. This shape of the transformation shows that the simple linear multicomponent analysis (as above) cannot be applied to this system, but that the nonlinear transformation must be taken into account. Our experiments have shown that for the reflection measurement in the skin within a wavelength range of 500–600 nm this transformation can be well approximated by a three-parametric hyperbola [14].

In this analysis not only the mixture of the two basic spectral components must be determined but also the parameter of the hyperbola:

$$H^{-1}(y) = (A\,y + C)/(F\,y + 1)$$

Starting the calculation with $C = F = 0$, the parameter is gradually changed until the best fit to the measured spectra is obtained. Again, the criterion of the minimum summed least squares of the errors is used. Our measurements reported so far were performed with the Rapidspektrometer of Kieler-Howaldwerke [15, 16].

In cooperation with the Sigma Company in West Berlin the Oxyscan was developed. This is a new, considerably smaller and easily transportable reflection oximeter which allows a clinical test of the method of hyperbola approximation using the basic spectral components. Figure 3 is a schematic drawing of

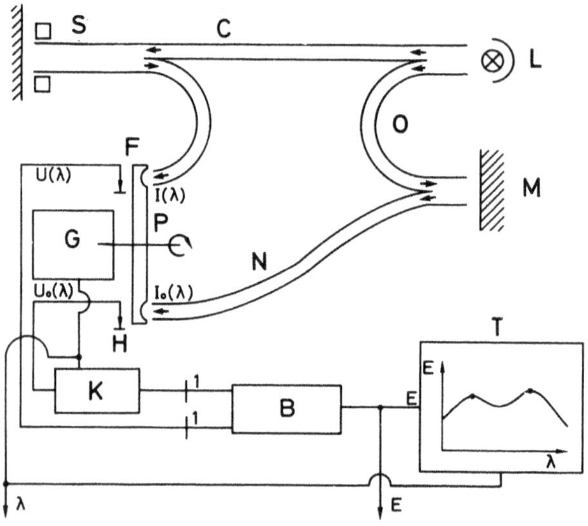

Fig. 3. Schematic drawing of the Oxyscan (Sigma, West Berlin). *B,* Logarithmic amplifier; *C,* flexible light guide (lamp-skin); *F,* photo cell (measuring light); *G,* motor; *H,* photo cell (reference light); *K,* signal processing; *L,* lamp; *M,* reference standard (MgO covered); *N,* light guide (reference standard–filter); *O,* light guide (lamp reference standard); *P,* filter disk; *S,* light guide to the skin, measuring head in contact with the skin; *T,* oscillographic screen. *E,* extinction; λ, wavelength; $U_0(\lambda)$, $U(\lambda)$, voltage of the photo cells; $I_0(\lambda)$, $I(\lambda)$, intensity of the reference and the measuring light

the Oxyscan. The spectral dispersion of light is accomplished by an interference filter disk rotating at a speed of approximately 30/s. The light emitted by the lamp penetrates a flexible light guide through the measuring head to the skin as well as to a white reference standard. From the measuring head a separated bundle of fibers guides the measuring light reflected from the skin towards the filter disk. The measuring light passes the filter and falls upon the measuring photo cell. The reference light guide is mounted in a fixed distance to the measuring light guide. The reference light falls upon the reference photo cell. The delay of impulses is compensated by signal processing. Then, the extinction is calculated and the extinction spectra immediately represented on the oscillographic screen. A parallel exit permits the processing of the signals in the computer.

To obtain an optimal signal-to-noise ratio the interference filter has a large spectral half-band width of approximately 15 nm. Since signals may be altered by a half-band width which is too large, and an interpretation without special nonlinear calibration curves is made impossible, test experiments of hemoglobin solution in cuvettes were therefore performed. These experiments resulted in a linear relation between extinction and concentration of HbO_2 under a wide range of measuring conditions. This proved that Oxyscan is suited for photometric measurements of oxygen saturation of hemoglobin [17].

In order to perform the hyperbola approximation with Oxyscan the spectra of all substances forming the reflection spectra must be known. Spectra of fully oxygenated and deoxygenated blood were determined as well as the zero line (spectra of the apparatus). Experiments on persons with different degrees of skin tan have shown evaluations to be more reliable considering an additional determination of the reflection spectra of hemoglobin-free skin. The spectra of hemoglobin-free skin can be obtained applying strong pressure to the skin surface by the measuring head, thus squeezing the blood from the outer skin vessels.

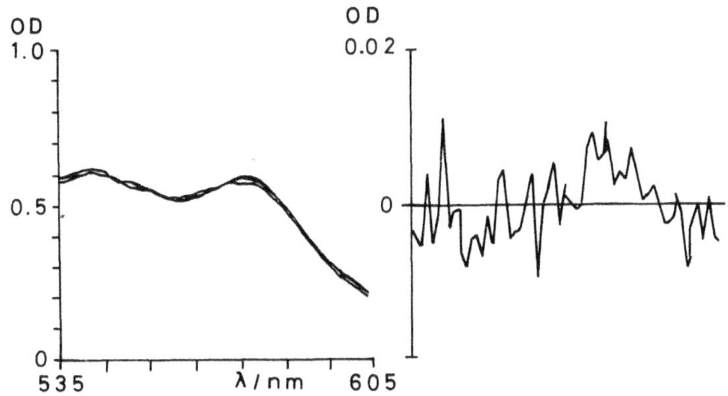

Fig. 4. Left, comparison of measured and recalculated skin reflection spectra. **Right,** the difference between the two spectra enlarged 25 times

Figure 4 (left) shows a reflection spectrum of skin and the corresponding recalculated spectrum using the hyperbola approximation method. It can be seen that the spectra are in good agreement. For better distinction the deviation between measured and calculated spectra is plotted 25-fold enlarged. The maximal deviation amounts to 0.01 OD (optical density); the average deviation is approximately half as much. Figure 5 gives another example of skin reflection spectra of the hyperemized forearm in a healthy adult; this displays the two-peaked oxygenated spectrum, the one-peaked deoxygenated spectrum, and the spectrum of the hemoglobin-free skin; on the right, the oxygenated spectrum corrected by the hyperbola approximation is shown. The saturation of the oxygenated blood amounted to 94.1% and that of the deoxygenated blood to 21.4%.

To prove the correctness of the hyperbola approximation method using basic spectra we used Oxyscan to determine the O_2 saturation in blood taken from the skin. To guarantee a sufficient O_2 saturation of the blood in all investigated vascular areas we provoked a skin hyperemia by application of Finalgon on the forearm. This spot was also treated with an infrared lamp for 10–15 min. During this period, O_2 breathing was applied for 10 min. After such a preparation, the noninvasively and the invasively measured O_2 saturation levels were undistinguishable [17]. These investigations show that it is possible to measure the oxygen saturation in the whole range between 0% and 100% within an accuracy of 5%.

The method has been applied as a control of the arterial oxygen saturation of skin blood in cases of disturbed circulation. It is especially well suited in controlling revascularization of skin flaps and replants. Figures 6 and 7 show some applications of hand surgery in the Chirurgische Universitätsklinik, Würzburg. Figure 6 gives the spectra of an axial pattern flap. Above on the left, the spectrum of the flap with normal blood supply is plotted (hemoglobin oxygenated to 65%). Pinching the pedicle after 1 week results in a totally deoxygenated spectrum. This demonstrates that no additional vascular supply has yet formed. After another week, total deoxygenation by pinching the flap is

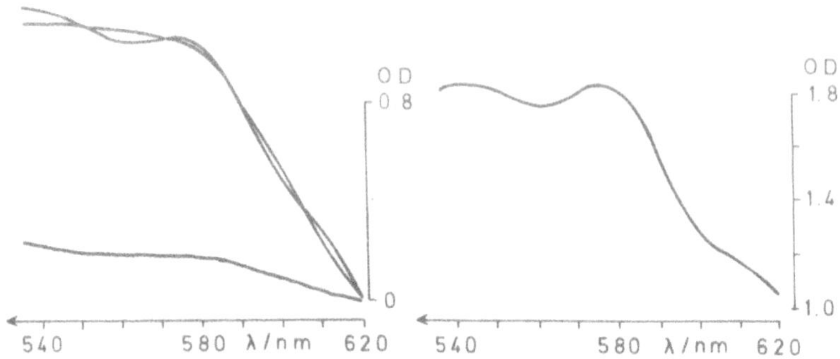

Fig. 5. Left, skin reflection spectra with partly oxygenated and deoxygenated blood and without blood. **Right,** the corresponding HbO_2 spectrum corrected by hyperbola approximation

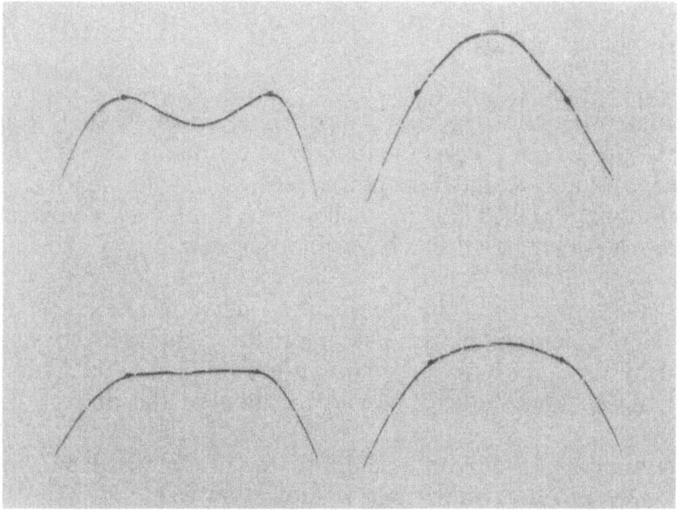

Fig. 6. Reflection spectra of an axial pattern flap. **Upper left,** spectrum with unclamped artery, 65% O_2 saturation. **Upper right,** 1 week after surgery with clamped artery, 0% O_2 saturation. **Lower,** 2 weeks after surgery with clamped artery; **left,** measured at the border zone, 35% O_2 saturation; **right,** measured at the zentral zone, 15% O_2 saturation

no longer possible. The saturation of hemoglobin in the periphery of the flap stays at 35% (bottom left) and in the center at 15% (bottom right). Figure 7 presents the spectra of a partially amputated big toe: top left, the spectrum of the stump (hemoglobin saturation 80%); top right, in comparison the spectrum of the other big toe (hemoglobin saturation 75%). The panel in the second row on the right presents the spectrum of the stump attached to the artery (hemoglobin saturation 60%). Pinching the artery results in the deoxygenated spectrum, as shown in the second row, left. After microsurgical replantation the spectrum remains deoxygenated with the tourniquet on. Restoring the circulation in this case produces a very good result: bottom right, a saturation of 80% is to be seen.

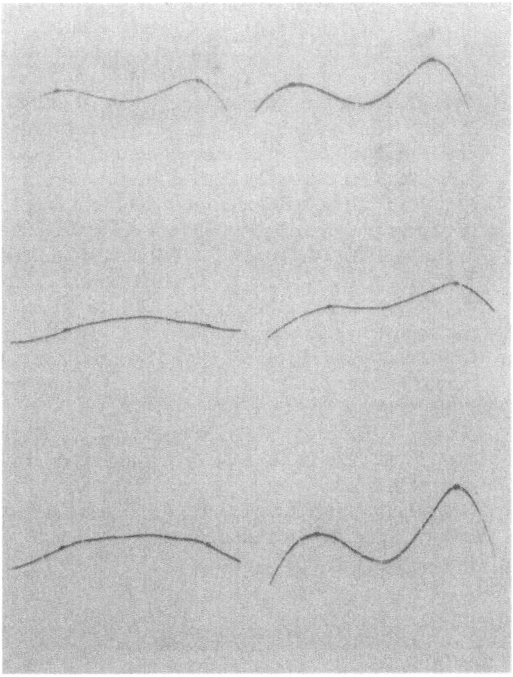

Fig. 7. Spectra of a freely transferred pulpa flap. **Upper right,** thumb of big toe, 80% O_2 saturation; **upper left,** other big toe (for comparison), 75% O_2 saturation. **Middle left,** section of big toe hanging on the artery, 60% O_2 saturation; **middle right,** tourniquet on, O_2 saturation 0%. **Lower right,** after microsurgical replantation (tourniquet still on), 0% O_2 saturation; **lower left,** tourniquet off, 80% O_2 saturation

These results show that the light-guide photometer Oxyscan provides an excellent noninvasive control of the state of oxygenation of skin flaps. Scanning the skin flap gives immediate local information. The instrument is easy to handle and needs no individual calibration.

References

1. Heilmeyer L (1933) Medizinische Spektrophotometrie. Ausgewählte Methoden und neuere Untersuchungsergebnisse an Körperfarbstoffen und Körperflüssigkeiten. Fischer, Jena
2. Kramer K (1934) Fortlaufende Registrierung der Sauerstoffsättigung im Blute an uneröffneten Blutgefäßen. Klin Wochenschr 13:379–380
3. Matthes K (1942) Untersuchungen über den Verlauf der Oxyhämoglobinreduktion in der menschlichen Haut. Pflugers Arch 246:70–91
4. Matthes K (1935) Untersuchungen über die Sauerstoffsättigungen des Arterienblutes. Arch Exp Pathol Pharmakol 179:698–711
5. Matthes K (1957) Kreislaufuntersuchungen am Menschen mit fortlaufend registrierenden Methoden. Thieme, Stuttgart
6. Millikan GA (1942) the oximeter, an instrument for measuring continously the oxygen saturation of arterial blood in man. Rev Sci Instrum 13:434–442
7. Nilsson NJ (1960) Oximetry. Physiol Rev 40:1–26
8. Wood EH (1950) Oximetry. In: Glasser O (ed) Medical physics, vol 2. Year Book Medical Publishers, Chicago, pp 664–680
9. Zijlstra WG (1953) Fundamentals and applications of clinical oximetry. Van Gorcum, Assen
10. Zijlstra WG, Mook GA (1962) Medical reflection photometry. Van Gorcum, Assen

11. Lübbers DW, Niesel W (1957) Über die Auswertung von spektralen Veränderungen in biologischen Objekten. Naturwissenschaften 3:59–60
12. Wodick R (1971) Neue Auswertverfahren für Reflexionsspektren und Spektren inhomogener Farbstoffverteilung, dargestellt am Beispiel von Hämoglobinspektren. Dissertation, University of Marburg
13. Wodick R, Lübbers DW (1973) Quantitative Analyse von Reflexionsspektren und anderen Spektren mit inhomogenen Lichtwegen an Mehrkomponentensystemen mit Hilfe der Queranalyse, II. Die Bestimmung von wellenlängenabhängigen Störungen und darin enthaltenen unbekannten Komponenten an Mehrkomponentensystemen mit Hilfe der Queranalyse. Hoppe Seylers Z Physiol Chem 354:916–922
14. Lübbers DW, Hoffmann J (1981) Absolute reflection photometry at organ surfaces. In: Kovách AGB, Monos E, Rubányi G (eds) Cardiovascular physiology. Heart, peripheral circulation and methodology. Akadémiai Kiadó, Budapest, pp 353–361 (Advances in physiological sciences, vol 8)
15. Lübbers DW, Niesel W (1959) Der Kurzzeitspektralanalysator: Ein schnell arbeitendes Spektralphotometer zur laufenden Messung von Absorption- bzw. Extinktionsspektren. Pflugers Arch 268:286–295
16. Niesel W, Lübbers DW, Schneewolf D, Richter J, Bötticher W (1964) Double beam spectrometer with 10 msec recording time. Rev Sci Instrum 35:578–581
17. Merschbrock U (1986) Ein neues Verfahren zur nichtinvasiven Bestimmung der intrakutanen Hämoglobinoxigenierung – Ein Beitrag zur Reflexionsoximetrie der Haut. Dissertation, University of Gießen

Innovative Aspects in Ventilation and Lung Function

Use of the VG Spectralab-M Quadrupole Mass Spectrometer in the Determination of Breath-by-Breath Gas Uptake

P. C. W. Beatty, B. Kay and T. E. J. Healy

Introduction

The mass spectrometer is potentially the most flexible and comprehensive gas concentration measurement instrument available to respiratory physicians and anaesthetists. From the early days of mass spectrometry, measurements of gas concentration in respired gases have been attempted [1], but it was not until 1964 when Fowler [2] reported his design for a magnetic sector machine for the measurement of respiratory gases that usable clinical mass spectrometers became available. With wider application of such machines, unacceptable errors of measurement were reported [3–5], especially when mass spectrometers were applied during anaesthesia. These errors were compounded in subsequent physiological measurements dependent on gas concentration measurement. Electronic correction of some errors [6, 7] produced acceptable accuracy for respiratory applications, but analysis [8] suggested that the obtained accuracy was insufficient for acceptable accuracy during anaesthesia.

The application of microprocessor technology to mass spectrometers intended for clinical use has now produced instruments which offer easy calibration procedures and accurate cracking pattern overlap correction even for anaesthetic application. This paper discusses the application of such a machine (VG Micromass Spectralab-M) to the measurement of breath-by-breath gas exchange. The Spectralab-M accuracy is discussed in the light of an accuracy standard derived from a range of physiological measurement. The algorithms required for online calculation of gas uptake during anaesthesia are shown. Finally, the hardware for a practical system based on the Spectralab-M is discussed.

Standard of Accuracy for Gas Concentration Measurements

Several authors have described accuracy standards for measurement of gas concentration, but must of these have been arbitrarily arrived at [9, 10] or based on surveys of exciting instruments whose accuracies have been found acceptable in clinical use [11]. An acceptable accuracy might be defined as the

Table 1. Error sensitivity factors for standard physiological measurements

Method	Gas	Parameter	Factor
Closed circuit measurement of oxygen uptake [13]	O_2	VO_2	0.76
	N_2		−1.15*
Closed circuit measurement of nitrous oxide uptake [14]	N_2O	VN_2O	−1.15 (systematic) 1.0 (random)
Closed circuit measurement of oxygen uptake and carbon dioxide output [15]	O_2	VO_2	0.00126
	CO_2	VCO_2	0.00077
Oxygen uptake by mass spectrometer and gas mixing box [16]	A_2	VO_2	1.0
	O_2		6.65
	N_2		19.54
	CO_2	VCO_2	1.0
Breath-by-breath gas exchange methods [17]			
I-E	CO_2	VCO_2	0.96
	O_2	VO_2	3.74
	N_2O	VN_2O	14.70
	N_2	VN_2	14.70
E	O_2	VO_2	5.30
	CO_2	VCO_2	−0.96*
I-EPS	O_2	VO_2	1.16
	CO_2	VCO_2	4.90
	N_2O	VN_2	21.80
EPS	O_2	VO_2	6.18
	CO_2	VCO_2	1.68
Gas exchange by gas titration methods [18]	CO_2	VCO_2	−5.81*
	O_2	VO_2	−5.81
Cardiac output by freon uptake [19]	N_2	Q_C	−1.38*
	Freon		11.16
Cardiac output by nitrous oxide absorption [20]	N_2O	Q_C	0.06
Cardiac output by acetylene absorption [21]	H_2O	Q_C	0.12
	C_2H_2		6.01
Respiratory exchange quotient [22]	CO_2	RQ	0.99
	O_2		4.14
	N_2		4.79
Blood gas concentrations	Halothane		1.00
Bubble equilibration and long inlet [23]	CO_2		0.04
Ventilation perfusion ratio [24]	CO_2		1.51

The table shows the effect of random error in the gas concentration measurements used in the calculation of the results at typical values quoted by the authors of the methods
* Negative signs imply that the actual value of the measurement decreases as the error increases. Where not otherwise stated the sensitivity factors refer to random errors.

maximum random error in original parameter that gives a physiologically insignificant variation in the measurement to be determined. The effect of an inaccuracy in one parameter in a calculation on the overall measurement accuracy may be studied by error sensitivity analysis. The error sensitivity factor (ESF) of a particular measurement is the ratio of the percentage error in the measured value to the percentage change in the parameter leading to that error:

$$\text{ESF} = \frac{V_{para}}{V_{res}} \qquad \frac{dV_{res}}{dV_{para}}$$

Thus, the effect on overall measurement error of a particular parameter error is given by multiplying the ESF for the parameter by the parameter error at the values of interest. To enable this to be done for a range of standard physiological tests, a small computer programme was written for a Sharp PC1500 pocket computer. Using the programme, 15 physiological measurements were examined. The results are summarised in Table 1, where the method of measurement, parameters determined, gas concentration measurements required and the overall ESF are noted. Where overall accuracy was determined by more than one gas concentration measurement, the method of propagation of errors [12] has been used to determine the overall effect on the measurement.

The overall errors in physiological measurements expected from an arbitary 0.1% vol/vol accuracy standard are below 3% of the overall measurement. The two worst cases are measurement of the breath-by-breath gas exchange using the I-EPS method as analysed by Swanson et al. [17]. This has a sensitivity to nitrous oxide concentration measurements of 21.8. The other case, again in the area of gas exchange, is the 19.54 sensitivity to nitrogen in the gas exchange measurements of Davies and Denison [16]. However, for a 0.1% vol/vol accuracy, these overall errors represent errors of less than 7.5 ml min^{-1} in a basal oxygen uptake of 250 ml min^{-1} and compare favourably with those described by Swanson et al. [17] for the errors induced by flow rate measurements in the same measurements.

Thus a 0.1% vol/vol standard for all gases appears to be an acceptable accuracy standard.

The Spectralab-M – Description

In many respects the Spectralab-M is a standard quadrupole mass spectrometer (Fig. 1). It has a two-stage heated stainless-steel capillary 1 m in length with a metrocil leak at the sample chamber. Ionisation is by electron bombardment. Ions are injected into the quadrupole and may be detected by an electron multiplier or a Faraday cup, depending on application. The pumping system is a turbo-molecular pump, backed by a rotary roughing pump. The ionisation control, peak locking and multiplier sensitivity are controlled by a seperate SX200 module which is itself under the control of an Apple IIe computer, integrated into the machine. Output of gas concentrations can be obtained either from an 8-bit D to A or RS232 interface. It is the software in the Apple that configures the machine, selects the gases to be examined and performs spectrum overlap eraser (SOE) and automatic sensitivity control (ASC) actions.

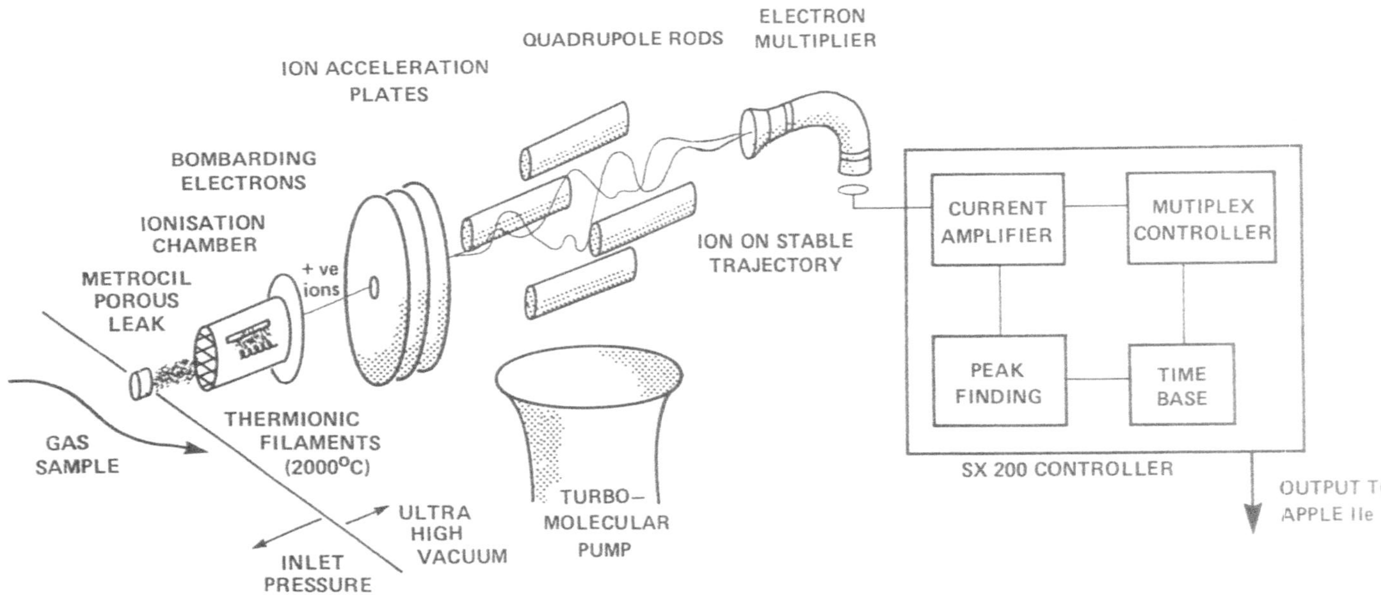

Fig. 1. Schematic of VG Micromass Spectralab-M mass spectrometer

Table 2. Configuration of the Spectralab-M used in test

Gas	M/Z ratio (amu)
O_2	32
N_2	28
A	40
N_2O	30
Isoflurane	149
Enflurane	67
CO_2	44

Table 3. Calibration mixture used in tests

Gas	Calibration point (% vol/vol)
O_2	10.87
A	0.44
N_2O	40.76
Isoflurane	1.40
Enflurane	1.41
CO_2	9.56
N_2	35.56

The Spectralab-M at Manchester has an optional cannulae inlet. Automatic calibration via a built-in calibration manifold is available as an additional option. This enables automatic peak split factors to be set up to give SOE action. Calibration in the Manchester machine uses a single-point technique where a mixture of known proportion is delivered from a gas mixture. Calibration takes less than 2 min per calibration gas.

Preliminary Accuracy Results

The static accuracy for the machine configured for seven gases have been measured (Table 2); the calibration mixture used is shown in Table 3. Comparison was made between the Spectralab-M and a Carl Zeis Jena interferometer. After calibration, two- and three-part mixtures, made using rotameters and a mixing box, were applied to the mass spectrometer. When stability had been reached, the concentration indicated by the interferometer and the mass spectrometer with the probe mounted immediately upstream of the mixing box were noted (Fig. 2). The result for oxygen is shown in Fig. 3. No alinearity was seen in any of the static calibration curves obtained, even where the calibration values in the mixture were well to one end of the range of interest. The overall

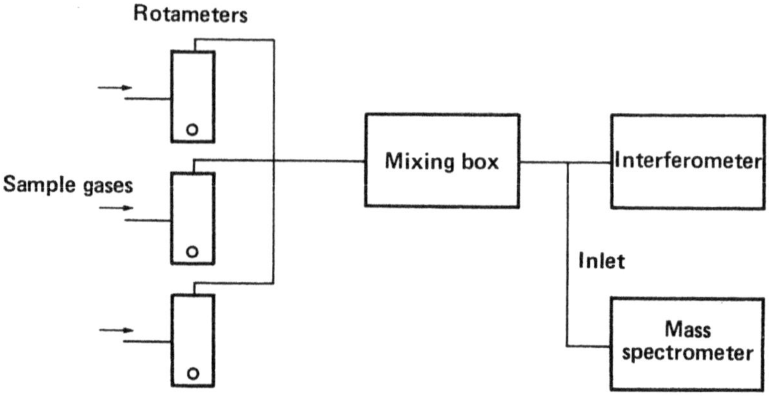

Fig. 2. Static calibration test equipment

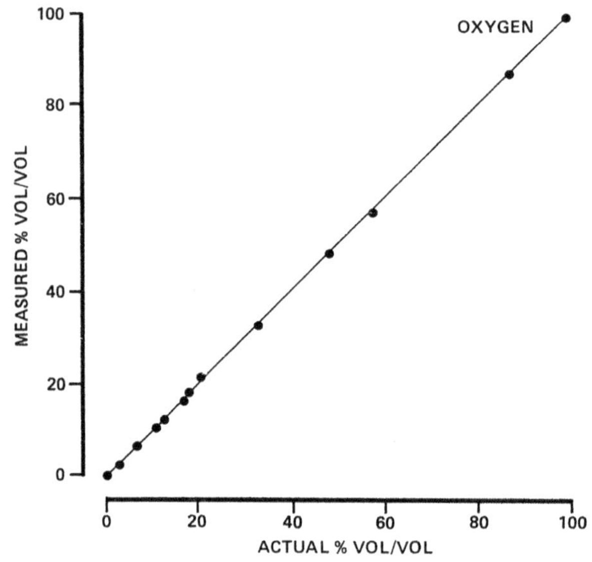

Fig. 3. Oxygen linearity curve shown in line of identity

confidence limits on repeatability and accuracy for the gases from these curves are shown in Table 4. No differences were seen regarding accuracy and repeatability.

Speed of response of the inlet was measured using the method of Gordon et al. [7]. The lag and 10%–90% rise times obtained are shown in Table 5.

From Table 4 it can be seen that the Spectralab-M does not meet the accuracy standard for all gases over the maximum possible dynamic range from 0% to 100% vol/vol, but in all cases for ranges around the chosen calibration points the accuracy is within the standard. This emphasises the importance of calibration near to envisaged operating point for any gas concentration measuring instrument.

Table 4. Confidence limits for part measurements: VG Spectralab-M compared with a Carl Zeis Jena interferometer

Gas	Range (% vol/vol)	Accuracy (± % vol/vol)
CO_2	0– 20	< ± 0.15
CO_2	0–100	< ± 0.38
O_2	0– 30	< ± 0.06
O_2	0–100	< ± 0.08
N_2	70– 90	< ± 0.22
N_2	0–100	< ± 0.22
N_2O	50–100	< ± 0.10
N_2O	0–100	< ± 0.21
A	0– 3	< ± 0.05
Enflurane	0– 5	< ± 0.08
Isoflurane	0– 5	< ± 0.05

Table 5. Response and lag times for Spectralab-M stainless steel inlet (flow rate 15 ml min^{-1})

		Rise		Fall	
Gases used	Gas monitored	Response time ms	Lag time ms	Response time ms	Lag time ms
O_2/Air	O_2	42	400	70	400
	N_2	60	400	70	400
N_2O/Air	N_2O	60	400	76	400
	O_2	42	400	40	400
10 % CO_2 in A/Air	A	45	500	88	500
	O_2	42	400	70	400
10 % CO_2 in A/N_2O	A	62	620	68	410
	N_2O	55	615	60	430
5 % Isoflurane in Air/N_2O	ISO	60	510	100	400
	N_2O	55	600	50	450
N_2O/Air [a]	N_2O	80	600	150	400
O_2/Air [a]	O_2	40	500	120	500

[a] Saturated with water vapour at 30°

Measurement of Breath-by-Breath Uptake

In measuring breath-by-breath uptake we have chosen to use two of the methods discussed by Swanson et al. [17], the I-E and I-EPS methods. In simple terms both compute the difference between the volume of a gas inspired and the volume of the gas expired at the mouth as follows:

$$V_x = F_{IX} V_I - F_{EX} V_E - S_X$$
$$S_x = F_{AX} V_L + V'_L F_{AX}$$
$$V_L = \frac{V_B}{F_{AB}} - \frac{F_{AB} \cdot V_L}{F_{AB}}$$

where

V_x = uptake on this breath of gas X
F_{IX} = mixed inspired concentration of gas X
V_I = inspired volume
F_{EX} = mixed expired concentration of gas X
V_E = expired volume
F_{AX} = alveolar concentration of gas X
V_L = change in lung volume from previous breath
V'_L = volume of end expiratory lung volume on previous breath
F_{AX} = change in alveolar concentration of gas X
F_{AB} = alveolar concentration of equilibrium gas B
V_B = apparent uptake of gas B

The difference between the methods lies in the correction for lung stores of gas between breath (S_X). In the I-E method S_X is assumed to be 0, i.e. lung stores of gas do not vary from breath to breath, and thus the gas uptake measured at the mouth represents alveolar uptake. In the I-EPS method one gas, which is in equillibrium with body tissues, and the respired mixture is used to examine changes in gas stores in lung volume changes. The apparent volume uptake of the gas may be used as a correcting factor for changes in lung gas stores. Continuous use of such a correction, as has been pointed out by Swanson [25], leads to compounded errors in the estimation of the lung volume. He has suggested the computation of effective lung value (ELV), which is that value of lung volume in the above equation that minimises between breath variation in apparent uptake. When using the I-EPS algorithm we intend to update the lung volume estimation in the equation above every 10 breaths to keep the breath-by-breath estimation of lung volume within reasonable bounds.

A schematic of the complete system to be used at Manchester is shown in Fig. 4. The mass spectrometer has been interfaced to a BBC microcomputer via the RS232 output from the Apple IIe. This interface carries the concentration information on all the gases in the selected order. Volume information passes into the microcomputer via the A to D convertor in the microcomputer. Inspired volume is available from one channel of the pneumotachometer and micromanometer; this need be provided only at the end of expiration since inspired concentration is assumed to be constant. However, the calculation of mixed inspired concentration requires instantaneous flow measurement during expiration.

After calculation, data is stored on floppy disc, and a real time display is generated by the microcomputer which gives a graphical representation of the most important uptake of interest and tabular information on inspired concentration, uptake, and ELV, updated every breath.

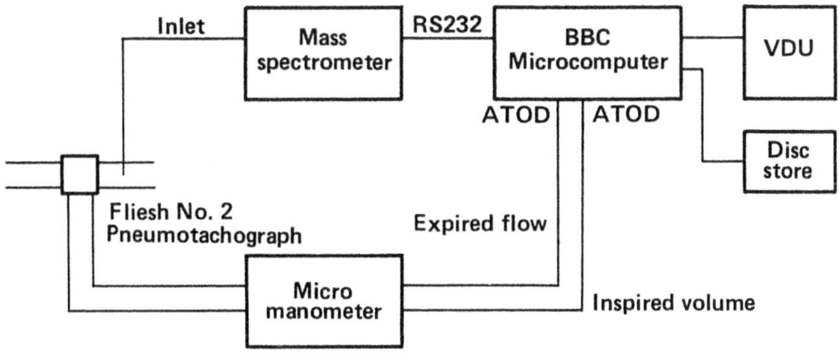

Fig. 4. Schematic of breath-by-breath system using Spectralab-M

Conclusion

The efficacy and, in many respects, the accuracy of the system envisaged by this paper for the measurement of breath-by-breath gas uptake relies on ease of use and calibration. The Spectalab-M is easy and quick to calibrate. The software to run the uptake calculation is EPROM based in the BBC microcomputer and automatically booted at switch-on. Calibration of the pneumotachograph system is required at reasonably regular intervals but should pose no problem to most clinicians. With such a system we hope that the full potential and flexibility of the mass spectrometer as a gas concentration measuring instrument can at last be realised in anaesthesia and intensive therapy.

References

1. Hunter JA, Stacey RW, Hitchcock FA (1949) A mass spectrometer for continuous gas analysis. Rev Sci Instrum 20:33
2. Fowler KT (1964) A mass spectrometer for the rapid and continuous analysis of gas mixtures in respiratory research. Thesis, London University
3. Salamonsen RF, Robinson GJ (1983) Computer correction of errors in anaesthesia monitoring by respiratory mass spectrometer. Comput Anaesth and Intens Care, Rotterdam
4. Graham JM, Salamonsen RF, Greer W, Kay B, Parkhouse J, Stephenson DK (1979) Errors in mass spectrometer monitoring during anaesthesia. IRCS Med Sci 7:5543
5. Graham JM, Salamonsen RF, Kay B, Greer W, Parkhouse J, Stephenson DK, Beatty P (1980) Mass spectrometry in anaesthesia: problems in using a mass spectrometer for monitoring anaesthetic gases and vapours. Acta Anaesthesiol Belg 31:265–278
6. Davies WOM, Spence AA (1979) A modification of the MGA 200 mass spectrometer to enable measurement of anaesthetic gas mixtures. Br J Anaesth 51:987–988
7. Gordon G, Boyd T, SalamonsenRF (1982) Remote monitored by mass spectrometer during anaesthesia. Br J Anaesth 54(1):15–22
8. Beatty PCW (1984) Potential inaccuracies in mass spectrometers with spectrum overlap erasure units used in anaesthesia. Clin Phys Physiol Meas 5(2):93–104

9. Fowler KT, Hugh-Jones P (1957) Mass spectrometry applied to clinical practice and research. Br Med J 1:1205–1211
10. Mosharaffa M (1970) Mass spectrometry in clinical and medical research. Res Dev 21:24–28
11. Fowler KT, Read J (1961) Cardiac oscillations in expired gas tension and regional pulmonary blood flow. J Appl Physiol 16:863
12. Barford NC (1969) Experimental measurements: precision, error and truth. Addson and Wesley, New York
13. Cohen A, Beatty PCW, Kay B, Healy TEJ (1984) Measurement of oxygen uptake: a method for use during nitrous oxide in oxygen anaesthesia. Eur J Anaesthesiol 1:63–75
14. Beatty PCW, Kay B, Healy TEJ (1984) Measurements of the rates of nitrous oxide uptake and nitrogen excretion in man. Br J Anaesth 56(3):223–232
15. Nunn JF, Pouliot JC (1962) The measurement of gaseous exchange during nitrous oxide anaesthesia. Br J Anaesth 34:752–763
16. Davis NJH, Dennison DM (1979) The measurement of metabolic gas exchange and minute volume by mass spectrometer alone. Respir Physiol 36:261
17. Swanson GD, Sodal IE, Reeves JT (1981) Sensitivity of breath-by-breath gas exchange measurements to expiratory flow errors. IEEE Trans Biomed Eng 28:749–754
18. Tsoi CM, Raemer DB, Westenskow DR (1982) Instrumentation simultaneously measuring VCO_2 and VO_2 in humans using titration methods. J Appl Physiol 52(3):786–791
19. Cotton EK (1976) The measurement of effective pulmonary blood flow in the newborn infant using a mass spectrometer. Crit Care Med (4):5
20. Lee G de J, Dubois AB (1955) Pulmonary capillary blood flow in man. J Clin Invest (34):1380–1390
21. Johnson RL, Sicer WS, Bishop JM, Forster RE (1960) Pulmonary capillary blood volume, flow and diffusing capacity during exercise. J Appl Physiol 15:893–902
22. West JB (1965) Ventilation/blood flow and gas exchange. Blackwell, Oxford, p 108
23. Mapleson WW, Willis BA, William B (1980) Blood-gas tension measurement in anaesthesia by bubble equilibration and mass spectrometry. Br J Anaesth 52:1061–1068
24. West JB (1965) Ventilation/blood flow and gas exchange. Blackwell, Oxford, p 107
25. Swanson GD (1980) Breath-to-breath considerations for gas exchange kinetics. Elsevier/-North-Holland, Amsterdam

Critical Oxygen Delivery Levels During Shock

N. S. Faithfull and S. M. Cain

Introduction

Whole body oxygen consumption (VO_2) is not usually supply-limited until oxygen transport (QO_2) is reduced below the critical oxygen delivery level (QO_2c) when the oxygen extraction coefficient (OEC) becomes maximal and VO_2 begins to decrease. QO_2c can be approached by lowering blood oxygen capacity by haemodilution or decreasing oxygen content of the arterial blood by hypoxia. This has been shown to result in similar levels of QO_2c, although the mixed venous partial pressures of oxygen (PvO_2 at QO_2c) were very different (Cain 1977). QO_2c may also be reached when cardiac output is reduced following acute haemorrhage (Schwarz et al. 1981; Becker et al. 1985).

Perfluorochemicals (PFCs) have a high solubility for respiratory gases and are available in emulsified form in the plasma substitute Fluosol-DA 20% (FDA), which is currently undergoing clinical trials in various countries. Because oxygen solubility of FDA is only 0.75 ml per 100 ml per 100 mmHg (Zander and Makowski, 1982; Grote et al. 1985), high inspired concentrations of oxygen (FIO_2) are usually administered concurrently.

PFC emulsions have low viscosity, particularly at low shear rates. This, combined with the very small particle size in the emulsion, may explain their ability to improve tissue oxygenation in the presence of ischaemic hypoxia during myocardial infarction (Biro 1983; Rude et al. 1982; Faithfull et al. 1985), cerebral infarction (Sutherland et al. 1984) and during perfusion of extremities following prolonged traumatic amputation (Smith et al. 1985). However, a short report exists (Erdmann 1982) that indicates that FDA may improve oxygen diffusion coefficients in the tissue, and this may also contribute to better oxygenation. If these effects are significant, haemodilution with FDA should lower QO_2c in the presence of a normal FIO_2. This paper reports experiments comparing QO_2c in FDA-haemodiluted animals under normoxia and hyperoxia with similar groups haemodiluted with dextran solutions.

Materials and Methods

Four groups of six mongrel dogs (12.3–23.1 kg body weight) were anaesthetised with sodium pentobarbitone 30 mg/kg; subsequently, about 30 mg per hour was given as indicated. The animals were intubated and ventilated to maintain end-expiratory carbon dioxide concentration at 4.5%–5%. Following surgical preparation suxamethonium was administered by intravenous drip.

Mixed expired gas was passed through a Harvard dry test gas meter, and samples were analysed for oxygen and carbon dioxide content. The output of the meter and the gas analysers were interfaced with a Cyborg ISAAC analog digital converter and processed by an Applie IIe microprocessor. Raw gas exchange data were displayed on a minute-to-minute basis and later corrected to standard temperature and pressure.

The right femoral artery was cannulated, and a catheter was placed in the pulmonary artery via the right external jugular vein. The pressures were transduced (Statham P23Db transducers) and displayed on a Beckman 5411B paper chart recorder. The heart rate was obtained from a Beckman type 9857 cardiotachometer coupler. The spleen was gently exposed via a midline laparotomy, and after the hylum had been ligated with multiple ligatures, the excision was closed with haemostatic clips.

In view of the frequent occurrence of hypotensive reactions to an initial dose of a fluorocarbon emulsion, all dogs were given a 2-ml intravenous dose of either FC-43 (an emulsion containing perfluorotributylamine) or FDA (containing a mixture of perfluorodecalin and perfluorotripropylamine). The protocol was not commenced until full recovery had taken place.

The animals were heparinised (1000 U/kg), and the experimental protocol began with a 20-min stabilisation period. Samples of femoral and pulmonary arterial blood were taken at 15 min. In all samples blood PO_2, PCO_2 and pH were measured at 37°C on a Radiometer BMS3 Mark 2 Blood Micro System or a Radiometer ABL 30 Acid/base Laboratory. Haemoglobin concentration and oxygen content were measured using an Instrumentation Laboratory Cooximeter 282, the output of which was corrected by reference to a calibrated Lexington Instruments Lex 02 Con-TL.

Haemodilution was then performed with either 6% dextran (molecular weight, 70 000) in Tyrode's solution or FDA by arterial bleeding and simultaneous intravenous administration of diluent. The volume required was calculated based on multiple previous haemadilution experiments performed in this laboratory. The target haemotocrit in these experiments was 25%.

After stabilisation, blood samples were taken at 45 min, and haemorrhage was commenced at 50 min. The initial bleed was of 5 ml per Kilogram body weight, and a calculated volume of red cells was administered to maintain the haematocrit near to 25%. The above steps were repeated every 20 min. The approach to QO_2c could be detected by the minute-to-minute oxygen consumption (VO_2) and bled volumes were reduced accordingly. The experiments were continued until the demise of the animals.

QO_2c for each animal was obtained by fitting two regression lines to the VO_2/QO_2 data using the least-squares method and determination of the intersection point (Mellits 1968). Statistical significance was assessed using analysis of variance, and statistical significance was accepted at the 5% level.

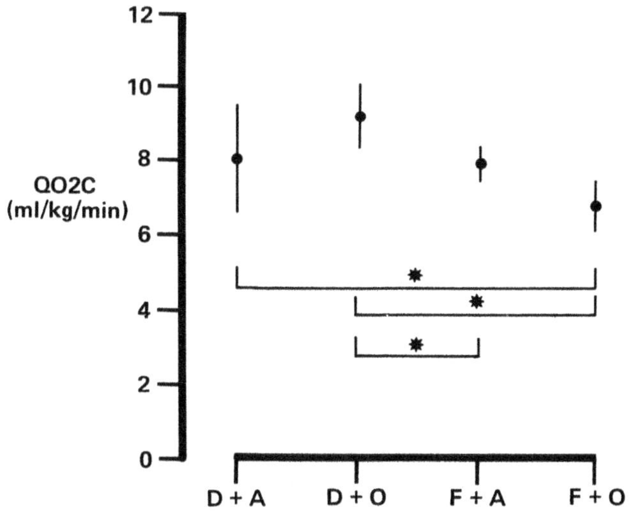

Fig. 1. Critical oxygen delivery levels *(QO_2C)* during haemorrhagic shock in dogs following haemodilution with 6% dextran or Fluosol-DA 20%. *D+A,* dextran and air; *F+A;* Fluosol and air; *D+O,* dextran and oxygen; *F+O,* Fluosol and oxygen. Statistical significance indicated by *asterisks* ($p < 0.05$)

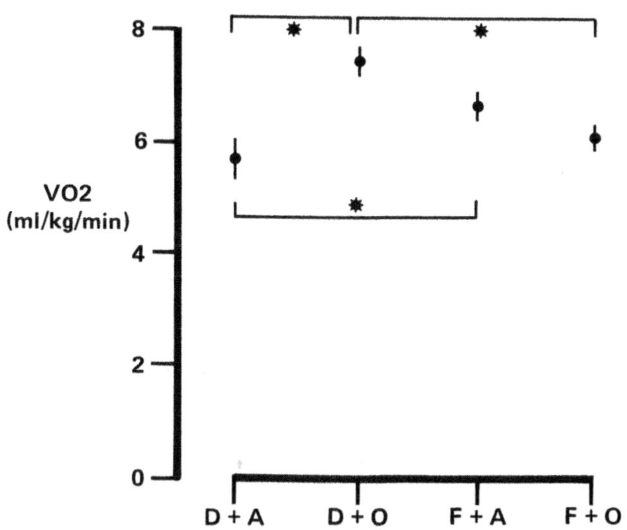

Fig. 2. Baseline values for oxygen consumption *(VO_2)*. *D+A,* dextran and air; *F+A,* Fluosol and air; *D+O,* dextran and oxygen; *F+O,* Fluosol and oxygen. Statistical significance indicated by *asterisks* ($p <$ 0.05)

Results

Calculated QO_2c values for each group are presented in Fig. 1. The lowest values were seen in the FDA group breathing 100% O_2 (F+O group). The values were significantly lower than the dextran group breathing air (D+A group) or that breathing 100% O^2 (D+O group). The values in the FDA group breathing air (F+A group) were significantly lower than the D+O group. However, there were significant differences in the baseline values for VO_2 (Fig. 2); these were estimated as the mean of all pre- and post-dilutional VO_2 values. Haemodilution had no significant effect on VO_2 in any group.

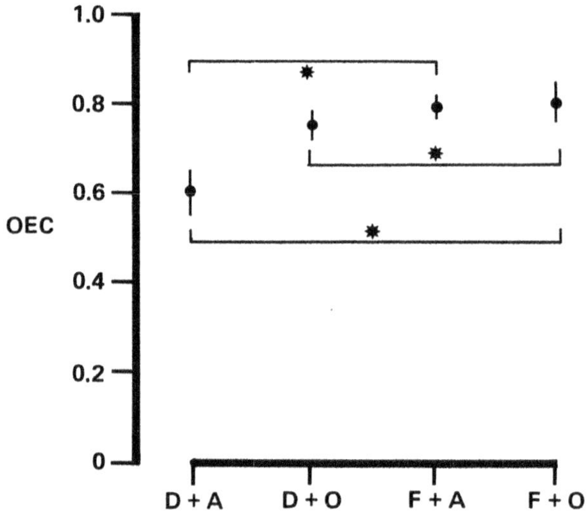

Fig. 3. Whole body oxygen extraction coefficients *(OEC)* at critical oxygen delivery levels during haemorrhagic shock in dogs following haemodilution with 6% dextran or Fluosol-DA 20%. *D+A,* Dextran and air; *F+A,* Fluosol and air; *D+O,* dextran and oxygen; *F + O,* Fluosol and oxygen. Statistical significance indicated by *asterisks* ($p < 0.05$)

QO_2c is partly dependent on the level of resting "plateau" VO_2; Fig. 3 shows values for oxygen extraction coefficients (OECs) at QO_2c. OEC is unaffected by VO_2 and hence can give a better overall impression of QO_2c levels when examining intergroup differences. OECs were significantly better in the F+O group than in the D+O and D+A groups. Interestingly, the values were significantly higher in the F+A group than in the D+A group in spite of the fact that under ambient air conditions FDA carries negligible amounts of oxygen.

The arterial oxygen tension (PaO_2) in the F+ O group during experimental protocol was 519.9 mmHg (± 11.5 SEM). Under these conditions the PFCs present in the blood would have been transporting significant quantities of oxygen and, because of the straight-line relationship between PaO_2 and content, the oxygen extraction from them (and also from the plasma) would have been over 90%. When all oxygen dissolved in plasma and PFCs is subtracted from the results, a figure for oxygen extraction from haemoglobin contained in the erythrocytes (HOEC) is obtained. HOEC at QO_2c was significantly greater in the F+A group than the F+O and D+O groups. The higher HOEC level in the F+A group is confirmed by examination of the values for mixed venous oxygen tension (PvO_2) at QO_2c. These values are shown in Fig. 4.

Discussion

The data presented above suggest that haemodilution with FDA, with or without supplemental oxygen, can result in greater oxygen extraction when QO_2 is limited and can hence reduce QO_2c. The explanation for this is probably multifactorial.

Haemodilution with FDA produces marked reduction in in vitro viscosity, particularly at low sheer rates (Naito and Yokoyama 1978), and at microcirculatory sheer rates this is less than when 6% dextran (molecular weight 70 000)

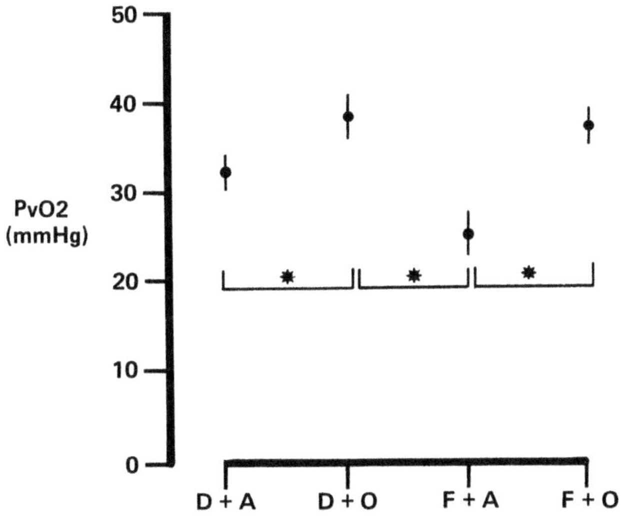

Fig. 4. Mixed venous oxygen tensions *(PvO₂)* at critical oxygen delivery levels during haemorrhagic shock in dogs following haemodilution with 6% dextran or Fluosol-DA 20%. *D+A*, Dextran and air; *F+A*, Fluosol and air; *D+O*, dextran and oxygen; *F+O*, Fluosol and oxygen. Statistical significance indicated by *asterisks* ($p < 0.05$)

is used as diluent (Biro 1982). Although the overall effect of haemodilution is a decrease in blood viscosity and total peripheral resistance (PR), Fan et al. (1980) demonstrated that increases in vascular hindrance may occur, particularly following haemodilution with dextran (Biro 1982). In this study significant falls in PR were seen in the F+A group following haemodilution but not in the D+A group.

Analysis of variance of post-dilutional PR levels revealed significantly lower values in the F + A group than in the D+A or D+O groups, while the F+O group had significantly lower PR than the D+O group. These results imply that dilution with FDA results in lower PR than corresponding dilution with dextran, and that hyperoxia, which is known to cause a decrease of capillary perfusion (Lindbom and Arfors 1984; Dulling 1972), may increase PR in the presence of both diluents. Indeed, hyperoxia may so seriously interfere with microcirculatory perfusion matching tat decreases in VO_2 may be seen when anaemic animals are ventilated with 100% O_2 (Chapler et al. 1984), although this was not observed in our experiments.

The deleterious effects of hyperoxia on microcirculatory oxygenation may well account for the lower extraction from haemoglobin-bound oxygen in the F+O and D+O groups. The effect of this on overall oxygen extraction was effectively masked in the F+O group by the substantial amounts of oxygen extracted from PFCs. Blood haemodiluted with FDA would have a greater oxygen-carrying capacity than blood at the same haematocrit level with dextran. This extra oxygen would contribute to oxygen consumption and result in a sparing effect on haemoglobin-bound oxygen. At maximum limitation of QO_2, oxygen extracted from PFCs accounted for 31.3% of VO_2 in the F+O group, whereas the corresponding figure in the F+A group was only 1.7%. The oxygen solubility of 6% dextran is 0.26 ml per 100 ml per 100 mmHg (Zander 1978), which differs little from that of plasma. Hence, in the D+A and D+O groups the oxygen delivery characteristics were almost identical to those of plasma-diluted blood. The occurrence of higher OECs in the F + O group than in the D + A or D + O groups is understandable, but the higher OEC in the F + A group than in the D+A group needs explaining.

Before haemoglobin-bound oxygen reaches the tissues it passes by diffusion through the red cell, across the red cell membrane, through plasma and through the capillary endothelial membrane. Traditional physiology (Krogh 1919) considers the capillary as a point source of oxygen and regards distance from the capillary as the primary variable determining tissue oxygen tension. This implies a small transcapillary gradient. However, Hellums (1977) has shown that resistance to mass transfer of oxygen is about the same in both blood and tissue, and recent mathematical analysis (Clark et al. 1985) has established the presence of a boundary layer near the red cell membrane acting as a significant barrier to oxygen transfer. Oxygen transfer across the capillary endothelium probably varies depending on the capillary bed concerned. For instance, the endothelium of the rete mirabile has a diffusion resistance 500 times that of plasma (Rasio and Goresky 1979), whereas the endothelium of lung and omentum do not appear to have diffusion resistance substantially greater than plasma (Sinha 1969).

Fluorocarbons easily traverse endothelial layers and are excreted through the lungs (Naito and Yokoyama 1978). Passage of pure or emulsified PFCs through the peritoneal membrane occurs (Clark et al. 1979) and Kolodgie et al. (1986) have identified fluorocarbon particles of various sizes within endothelial cells, interstitium and necrotic myocytes in the necrotic zone of myocardial infarction following treatment with FDA.

If, indeed, FDA improves oxygen extraction by improving diffusion through the red cell boundary layer, the endothelium or the tissues, higher tissue oxygen tensions should occur after administration of fluorocarbons. Following administration of 100% oxygen to dogs that had received either FDA or dextran, dramatic 285% increases in myocardial oxygen tensions were measured in the FDA group (Rude et al. 1984). In the dextran group the increases amounted to a mean of 136%. Conversely, higher oxygen extractions would lower the venous partial pressures of oxygen, as we saw in our experiments.

In conclusion, the present study has demonstrated that haemodilution with FDA, with or without additional oxygen, can result in significant increases in whole body oxygen extraction as compared to animals similarly diluted with 6% dextran in Tyrode's solution. This effect is not only due to the superior oxygen transport afforded by PFCs; indeed, hyperoxia detrimentally effects oxygen extraction from the red cell. When breathing ambient air, the effects may be caused by a combination of the superior rheological properties of FDA and a postulated diffusion facilitation phenomenon, leading to decreases in critical oxygen delivery levels during haemorrhagic shock.

Summary

Haemodilution with Fluosol-DA 20% (FDA), an oxygen-carrying plasma substitute containting fluorocarbons, has the ability to improve tissue oxygenation in conditions of ischaemic hypoxia. To determine the effect of FDA on critical oxygen delivery level (QO_2c), two groups of six phenobarbitone-anaesthetised, air-ventilated, splenic-clamped mongrel dogs were haemodiluted to a haematocrit of 25% with FDA or 6% dextran solution. Online oxygen consumption (VO_2) was obtained from expired oxygen concentrations and volumes and whole body oxygen flux (QO_2) was calculated from mixed venous and arterial oxygen contents and the Fick derived cardiac output. Two further groups were similarly haemodiluted and ventilated with 100% O_2; VO_2 in these groups was obtained using a spirometer and carbon dioxide absorption. QO_2 was decreased progressively by bleeding in 1.5- to 2.5 ml per kilogram steps until the demise of the animals. QO_2 was estimated at each step and haematocrit was kept at 25% using packed cells. VO_2/QO_2 pairs were calculated at each step, and QO_2c was determined for each animal by fitting two lines through the data, obtained using the least-squares method. Statistical analysis was assessed using an analysis of variance.

QO_2c was significantly less in the FDA + oxygen (F+O) group than in either the dextran + oxygen (D+O) or the dextran + air (D+A) group. Analysis of oxygen extraction at QO_2c, which effectively normalised results for differences in resting VO_2, revealed that the FDA + air group (F+A) had significantly better extraction than the D + A group. When fluorocarbon and plasma-dissolved oxygen was subtracted, the oxygen extraction in the F + A group was significantly better than in the D + A and F + O groups. The results imply that FDA haemodilution while respiring air can improve oxygen delivery from the red cell. Breathing 100% O_2 interferes with this process.

References

Becker L, Schumacker P, Nelson DP, Saltz SA, Rowland J, Long GR, Wood LDH (1985) Influence of FIO_2 on the relationship between oxygen delivery and uptake (VO_2) in the dog. Fed Proc 44:809

Biro GP (1982) Comparison of acute cardiovascular effects and oxygen-supply following haemodilution with dextran, stroma-free HB solution and FC suspension. Cardiovasc Res 16:194–204

Biro GP (1983) Fluorocarbon and dextran hemodilution in myocardial ischaemia. Can J Surg 26:163–168

Cain SM (1977) Oxygen delivery and uptake in dogs during anemic and hypoxic hypoxia. J Appl Physiol 42:228–234

Chapler CK, Cain SM, Stainsby WN (1984) The effects of hyperoxia on oxygen uptake during acute anemia. Can J Physiol Pharmacol 62:809–814

Clark A, Federspiel WJ, Clark PAA, Cokelet GR (1985) Oxygen delivery from red cells. Biophys J 47:171–181

Clark LC, Moore RE, Diver S, Miller ML (1979) A new look at the vapour pressure problem in red cell substitutes. In: Proceedings of the 4 th international symposium on perfluorochemical blood substitutes. Excerpta Medica, Amsterdam, pp 55–67

Dulling BR (1972)Microvascular responses to alterations in oxygen tension. Circ Res 31:481–489

Erdmann W (1982) O_2 diffusion coefficients. In: Frey R, Beisbarth H, Stossek K (eds) Oxygen carrying colloidal blood substitutes. Zuckschwerdt, Munich, p 143

Faithfull NS, Erdmann W, Fennema M (1985) Oxygen tensions in the ischaemic myocardium following haemodilution with fluorocarbons or dextran. Br J Anaesth 57:816–817P

Fan F-C, Chen RYZ, Schluessler GB, Chien S (1980) Effects of hematocrit variations on regional haemodynamics and oxygen transport in the dog. Am J Physiol 238:H545–H552
Grote J, Steuer K, Muller R, Sontgerath C, Zimmer K (1985) O_2 and CO_2 solubility of the fluorocarbon emulsion Fluosol-DA 20% and O_2 and CO_2 dissociation curves of blood – Fluosol-DA 20% mixtures. Adv Exp Med Biol 191:453–461
Hellums JD (1977) The resistance to oxygen transport in the capillaries relative to that in the surrounding tissue. Microvasc Res 13:131–136
Kolodgie FD, Dawson AK, Forman MB, Virmani R (1986) Effect of perfluorochemical (Fluosol-DA) on infarct morphology in dogs with permanent coronary artery occlusion. Virchows Arch [B] 119–134
Krogh A (1919) The number and distribution of capillaries in muscles with calculations of the oxygen partial pressure head necessary for supplying the tissue. J Physiol 52:409–415
Lindbom L, Arfors K-E (1984) Non-homogeneous blood flow distribution in the rabbit tenuissimus muscle. Acta Physiol Scand 122:225–233
Mellits DE (1968) Statistical methods. In: Cheek DB (ed) Human growth. Lea and Febiger, Philadelphia, pp 19–38
Naito R, Yokoyama K (1978) Perfluorochemical blood substitutes. Green Cross, Osaka (Technical information series, no 6)
Rasio EA, Goresky CA (1979) Capillary limitation of oxygen diffusion in the isolated rete mirabilis of the eel (*Anguilla anguillaris*). Circ Res 44:498–503
Rude RE, Golgar D, Khuri SF, Kloner RA, Karaffa S, Muller E, Clark LC, Braunwald E (1982) Effect of intravenous fluorocarbons during and without O_2 enhancement on acute myocardial ischemic injury assessed by measurement of intramyocardial gas tensions. Am Heart J 103:986–995
Rude RE, Bush LR, Tilton GD (1984) Effects of fluorocarbons with and without oxygen supplementation on cardiac hemodynamics and energetics. Am J Cardiol 54:80–883
Schwarz S, Frantz RA, Shoemaker WC (1981) Sequential hemodynamic and oxygen transport responses in hypovolemia, anemia and hypoxia. J Appl Physiol 10(241):H864–H871
Sinha AK (1969) Oxygen uptake and release by red cells through capillary wall and plasma layer. Thesis, University of California, San Francisco
Smith AR, van Alphen W, Faithfull NS, Fennema M (1985) Limb preservation in replantation surgery. J Plast Reconstr Surg 75:227–237
Sutherland GR, Farrar MK, Peerless SJ (1984) Effect of Fluosol-DA on oxygen availability in focal cerebral ischemia. Stroke 15:829–835
Zander R (1978) Oxygen transport by solutions for blood replacement in comparison with other infusion solutions. Klin Wochenschr 56:567–573
Zander R, Makowski HV (1982) Life without haemoglobin? In: Frey R, Beisbarth H, Stossek K (eds) Oxygen carrying colloidal blood substitutes. Zuckschwerdt, Munich, pp 133–141

Highly Efficient Jet Ventilation

W. L. DEN DUNNEN AND T. MOSTERT

Introduction

At the 1984 symposium "Anaesthesia–innovations in management" we described the method of pneumatic controlled circulation [1]. In this technique a jet insufflation of the ventilatory gases is used which must technique a jet insufflation of the ventilatory gases is used which must be adjusted synchronously with the pressure in the capillary meshwork around the alveoli, providing a maximum difference between pneumatic and hydraulic pressure in the alveoli. In this ventilation technique the pneumatic control is adjusted in such a way that the momentary blood pressure in the capillary meshwork determines O_2 and CO_2 diffusion. The ventilation is adjusted in such a way that the required gas exchange corresponds to the momentary transmembranous pressure. In the alveoli the transmembranous pressure is zero. There are only minor pressure fluctuations–otherwise the membranes would be damaged.

The interaction of pneumatics and hydraulics in the thoracic cavity is represented in Fig. 1. These triple communicating vessels show the individual compliances leading to the total thoracic compliance. The displacement of the pistons can be used to indicate individual compliance. This model was thoroughly described in the proceedings of the symposium "Perspectives in high-frequency ventilation" held at Rotterdam in 1983 [2]. Compartment A_1 in Fig. 1 represents the airway tract; compartment A_3 represents the vascular tract, especially the highly compliant capillaries and veins in the lungs. When piston A_1 is pushed downward with high dP/dt (jet insufflation), piston A_3 goes up more than piston A_2, which represents thorax elasticity. The mass of the thorax itself makes it impossible to distend so suddenly. Figure 1B shows the situation in which piston B_1 is pressed upward by the force of the blood mass in the capillaries around the alveoli. Figure 1C shows the situation when gases are insufflated at too-high pressures with too-high rate of change of pressure (dP/dt). The effect on the circulation of such jet ventilation can be easily measured during asynchronous jet ventilation in human beings as well as in animal experiments. We observed harmonic oscillations in the pulmonary artery wave curve, indicating a direct effect on the circulation [1].

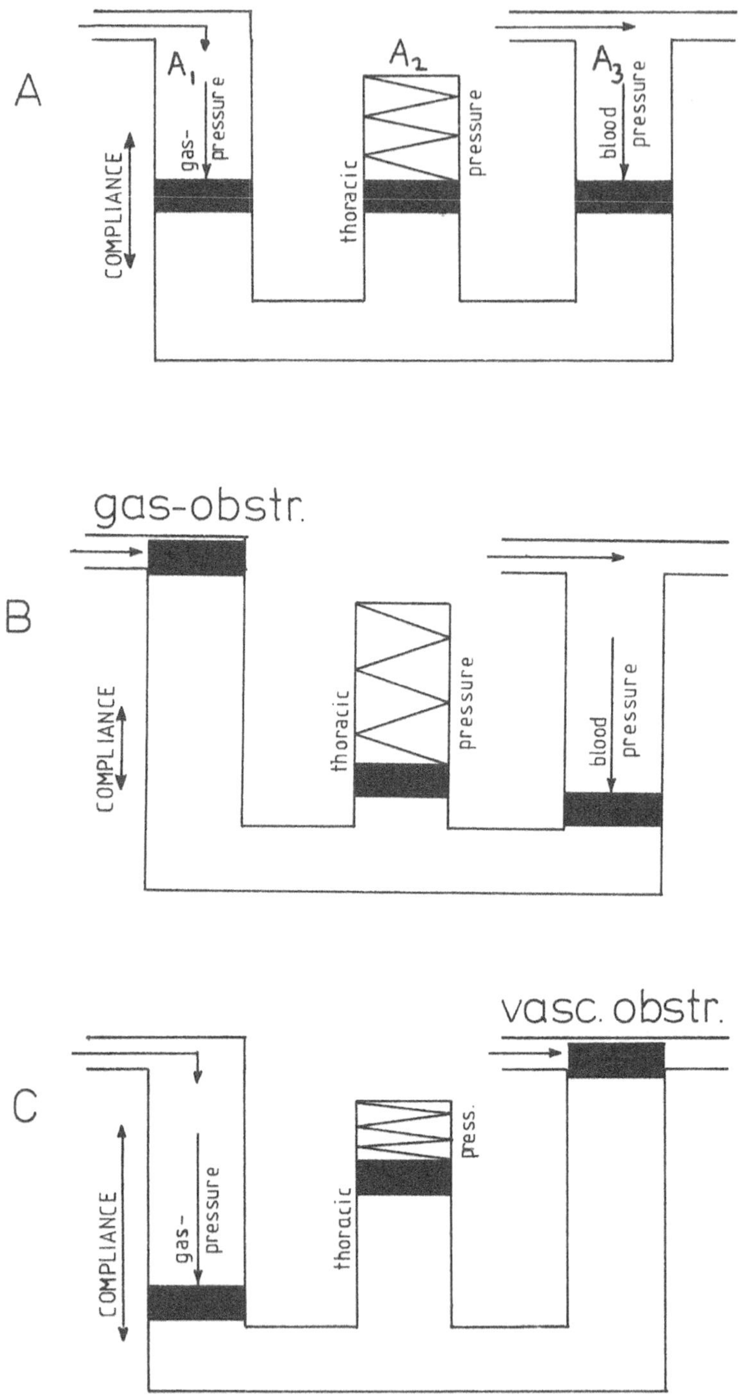

Fig. 1A–C. A model for "total thoracic compliance"

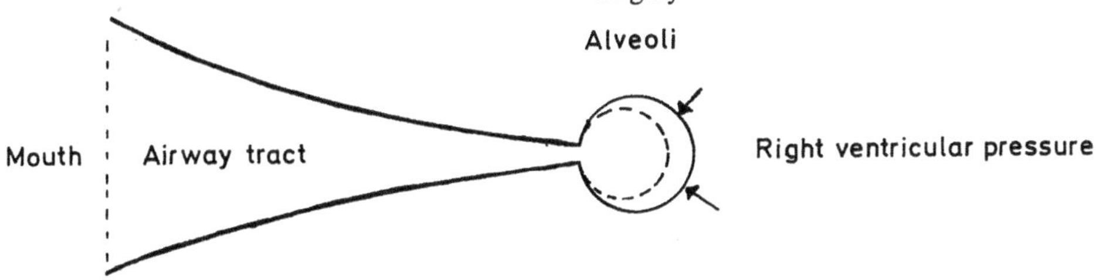

Fig. 2. A model of the lung

Based on this interaction we postulated in 1984 that there exists a natural influence of the blood entering the lung capillaries on the alveolar gas exchange. The blood mass in the capillaries causes the alveoli to be partially emptied during the right ventricular systole. Based on our findings we subdivided lung ventilation into: (a) exogenous respiration, the already known respiration which mainly ventilates the larger airways, and (b) endogenous respiration, the alveolar ventilation caused by right ventricular ejection. The endogenous respiration we named the cardioalveolar respiration pump (CARP).

Figure 2 shows a lung model. When gases enter the lung the resistance for gas molecules increases because of the smaller diameter of the channels. The connection to the alveoli are the bronchiolar and alveolar ducts, which have the highest resistance. In our opinion the "open trumpet" model cannot be accepted in lung physiology when a high d*P*/d*t* of the ventilatory gases is used because the individual diameters of the airway channels cannot be summarized, ignoring the impedance characteristics of the airway tract. The impedance characteristics change with the d*P*/d*t* of the ventilatory gases. (Summarizing the resistance of the individual airway channels of the same order would ignore physical laws.) However, in our opinion there is no relevant alveolar gas exchange based on Tailor's diffusion; since Tailor's diffusion is but a minor part of the nonturbulent gas mixing, it can be ignored.

We should realize that the lung is not a static but rather a very dynamic system. Because the mass of the thorax is relatively large, the thorax is not able to move in resonance with the endogenous respiration and thus moves more slowly. Based on the hypothesis of the CARP, we began ventilating patients synchronously with the heart duty cycle in such a way that the (alveolar) gas pressure falls during right ventricular systole and rises during right ventricular diastole. We tried to reduce the insufflation pressure and volume to such an extent that no influence on the pulmonary artery wave curve was observed. We refer to this kind of artificial ventilation as being in resonance with the CARP.

Methods

Insufflation of the gases in the trachea is performed through a 10-F suction catheter with one opening on the tip (Unoplast, Denmark). The catheter is placed in a routine endotracheal tube (8–9 mm internal diameter) or a Mallinckrodt Hi-Lo endotracheal tube. The tip lies halfway down the tube. For monitoring endotracheal pressure and temperature, a 7-gauge Swan-Ganz thermodilution (TD) catheter is used. The catheter is modified by shortening the tip just distally to the thermistor. The tip of the catheter is positioned about 2 cm above the carina. The proximal and insufflation ports of the TD catheter are permanently closed. The insufflation and Swan-Ganz catheters are fixed into a special T-piece (Bronchoscop-Aid, Dryden, USA).

Figure 3 shows the endotracheal tube together with the Bronchoscop-Aid. The expiratory port is closed with a one-way valve to prevent entrainment of room air during insufflation, and minute volume is monitored through this port using a Wright respirometer. The procedure to determine the exact ventilator settings is as follows. The intervals of the cardiac duty cycle are measured from the top of one ECG-QRS complex to the next and indicated in milliseconds. Then, a decision is made to divide the cardiac cycle into an inspiratory and expiratory phases in such a way that the total time of the ventilator setting never exceeds the time of one cardiac duty cycle. An asynchronous jet ventilation run is then made in such a way that the insufflation rate almost equals the heart rate. The inspiratory volume is limited to approximately 150–200 ml to prevent any risk of barotrauma. Doing so, and obvious harmonic oscillation is found in the oesophageal pressure, pulmonary artery pressure and central venous pressure.

A short run is recorded at 25 or 50 mm/s. The time in milliseconds is then measured from the top of the ECG QRS complex to the top of the pulmonary artery wave curve in the interval in which the greatest amplitude can be observed. It is also possible to measure the interval from the ECG QRS complex of the cardiac cycle in which the smallest pressure fluctuation in the oesophagus can be observed. Figure 4 shows the principle of this calculation. In this recording the oesophageal pressure fluctutation is used to calculate the correct setting. The horizontal arrow in Fig. 4 shows the cardiac cycle where the optimal setting can be found. One must realize that minor pressure fluctuations in the oesophagus coincide with an optimal synchronization of the heart action. The hydraulic and pneumatic systems aid each other, and minor transthoracic energy is dissipated. The time determined is used to set the trigger delay of the jet ventilator, starting at the ECG QRS complex. The delay time coincides the ventricular systole. The rest of the time of the cardiac duty cycle can be used for the insufflation of the ventilatory gases. Normally an insufflation time of 300–400 ms is enough to insufflate 70–100 ml gas. The insufflation coincides the ventricular diastole.

During the experiments the physical properties of the Mallinckrodt Hi-Lo tube were tested.

The Ventilator. Our own designed (jet) ventilator was used as described in the proceedings of the 1984 symposium referred to above [1]. The electronic settings used for this research were as follows.

Heart-synchronous mode. Insufflation pulses from about 20 ms to 1 s can be set. A delay can be set using the patient's ECG as a time reference. The jet valve can be triggered by an external trigger or as a function derived from the

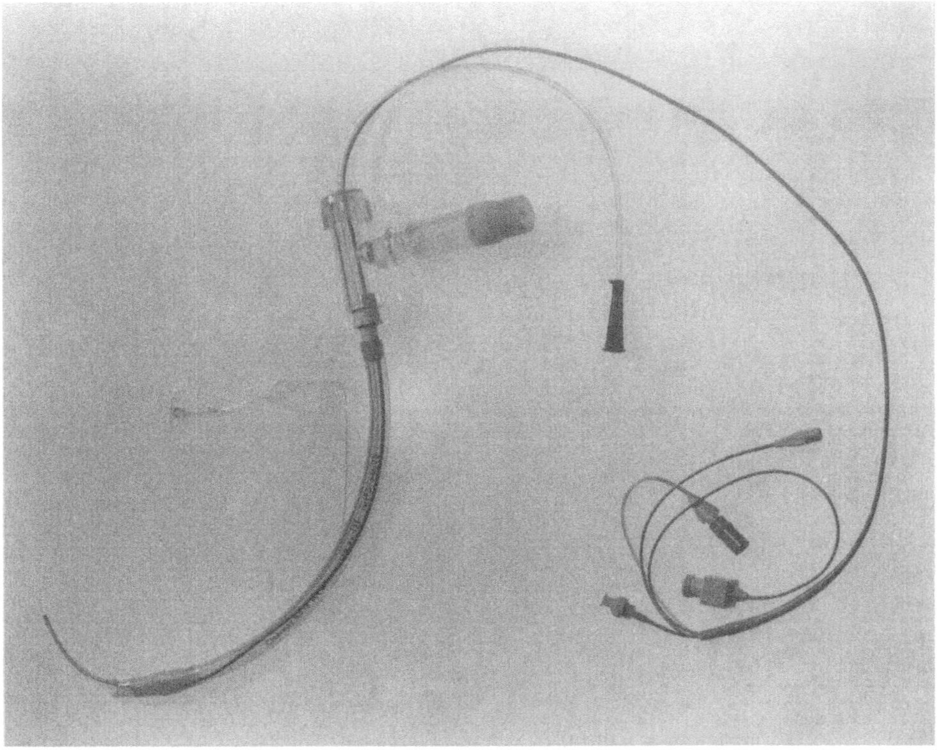

Fig. 3. Photograph of endotracheal tube and Bronchoscop-Aid (Dryden, USA)

pulmonary artery pressure and the ECG. During the experiments we preferred digital hand setting although we have developed microprocessor programmes to calculate the control function.

Anaesthetic Technique. All patients were scheduled for routine abdominal surgery. Premedication was administered 2 h before induction: 2.5 mg lorazepam (Temesta). They were given an intramuscular dose of 5 mg dehydrobenzperidol (droperidol) and 0.1 mg fentanyl 45 min before induction. Normovolaemia was ensured by pre-operative infusion of Ringer's lactate solution. Anaesthesia was induced with etomidate (Hypnomidate 0.2 mg/kg body wt.), and after muscle relaxation had been produced by a dose of suxamethonium (1 mg/kg body wt.), the patients were intubated with a disposable endotracheal tube or the special Mallinckrodt Hi-Lo endotracheal tube (8–9 mm internal diameter). These tubes have an supplementary channel for jet insufflation and pressure measurement. Anaesthesia was maintained with 33% oxygen and 66% nitrous oxide, supplemented with neuroleptanaesthesia using dehydrobenzperidol and intermittent doses of fentanyl. Muscle relaxation was obtained using alcuronium (0.2 mg/kg body wt.), and the patients were ventilated with a Dräger Spiromat ventilator to achieve an end-expiratory carbon dioxide level of between 3 and 4 vol.%.

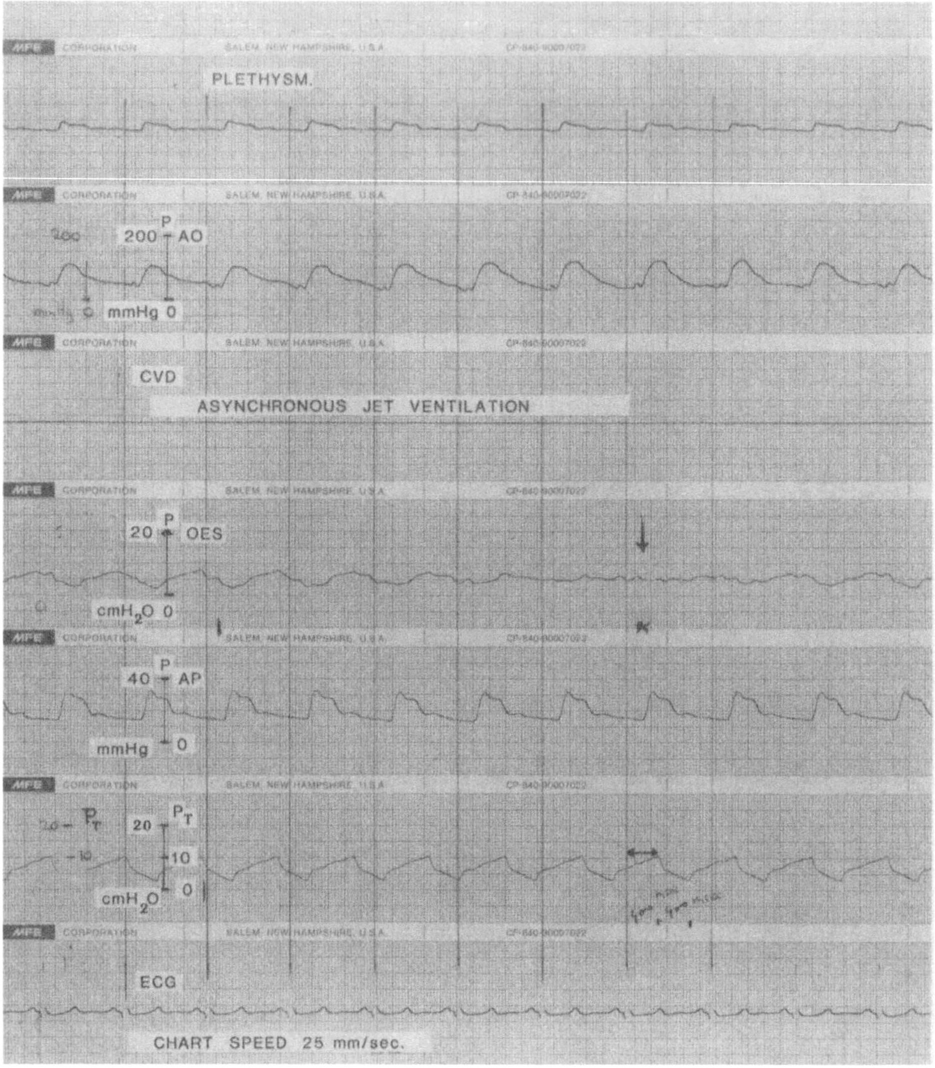

Fig. 4. Measurements for use of oesophageal pressure fluctuation in determining exact ventilator settings. *Horizontal arrow*, the cardiac cycle for optimal setting

The jet ventilation started after normalization of the circulatory volume. Infusion therapy was continued until pre-operative blood pressure was re-obtained. During steady-state anaesthesia runs were made with the jet valve ventilator using the same oxygen concentration of the ventilatory gases. Initially we made runs of a few minutes; later we continued this type of ventilation for periods of longer than 2 h.

Monitoring. Intra-arterial pressure monitoring was performed in all patients, using an 18-gauge intra-arterial catheter, introduced into the radial artery. A 7-

gauge Edwards Swan-Ganz catheter was inserted using a Cordis catheter introducer system via the cubital vein. A 14-F "national catheter" gastric tube with an oesophagus balloon was inserted. The stomach was emptied, and the gastric tube was closed off. This enabled monitoring of the intra-oesophageal pressure to be performed via the oesophageal balloon. The intravascular catheters and the balloon of the oesophagus catheter were connected to Gould pressure transducers (type P231D) and the intratracheal pressure Swan-Ganz catheter was connected to an Ailtech microtransducer (type MS20EA44ABS). A Siemens or Dräger capnograph was used during IPPV. ECG and pressure monitoring was performed with Hewlett-Packard devices (78000 series). An eight-channel MFE strip-chart recorder was used.

Results

Figure 5 shows a recording in which the setting was continued for more than 2 h. The inspiratory tidal volume was 110 ml. The intratracheal pressure (lowest trace) varied between +3 and +9 cm H_2O only. Arterial pCO_2, pO_2 andO_2 saturation levels are presented in Table 1. All other patients showed the same results although minor respiratory volumes of about 100 ml were used at heart frequency.

Figure 6 shows a recording in which the intratracheal pressure was measured through the pressure opening of the Mallinckrodt Hi-Lo tube and via the tip of the Swan-Ganz catheter, which was inserted more distally in the trachea. The second trace from the bottom shows the intratracheal pressure. At the left the pressure fluctuations were measured through the Swan-Ganz catheter tip opening. The right-hand part this trace is after changing to the Mallinckrodt pressure lumen. An obvious difference can be observed. The intratracheal pressure fluctuations measured via the Mallinckrodt tube are smaller than those measured more distally in the trachea.

Discussion

Using a suction catheter as a jet cannula has the advantage that it always positions itself in the middle of the trachea, but the opening of the jet lumen of the Mallinckrodt tube is fixed on one side with its opening directed causing more turbulence. Thus, we found the best jet properties with the 10-F suction catheter and the best pressure measurements by the distal opening of the Swan-Ganz catheter. The oesophageal pressure provides a reliable method of monitoring the effects of jet ventilation on the pulmonary circulation. As soon as harmonic oscillations occur, one should realize that the patient's lung circulation is impaired during the periods that great pressure fluctuations occur. This is not found when a thermodilution method is used to monitor the cardiac output. The monitoring system should be air filled because of the changing characteristics of the pressure-monitoring system when high dP/dt of the gases is used. The gain of the pressure amplifier should be at maximum level.

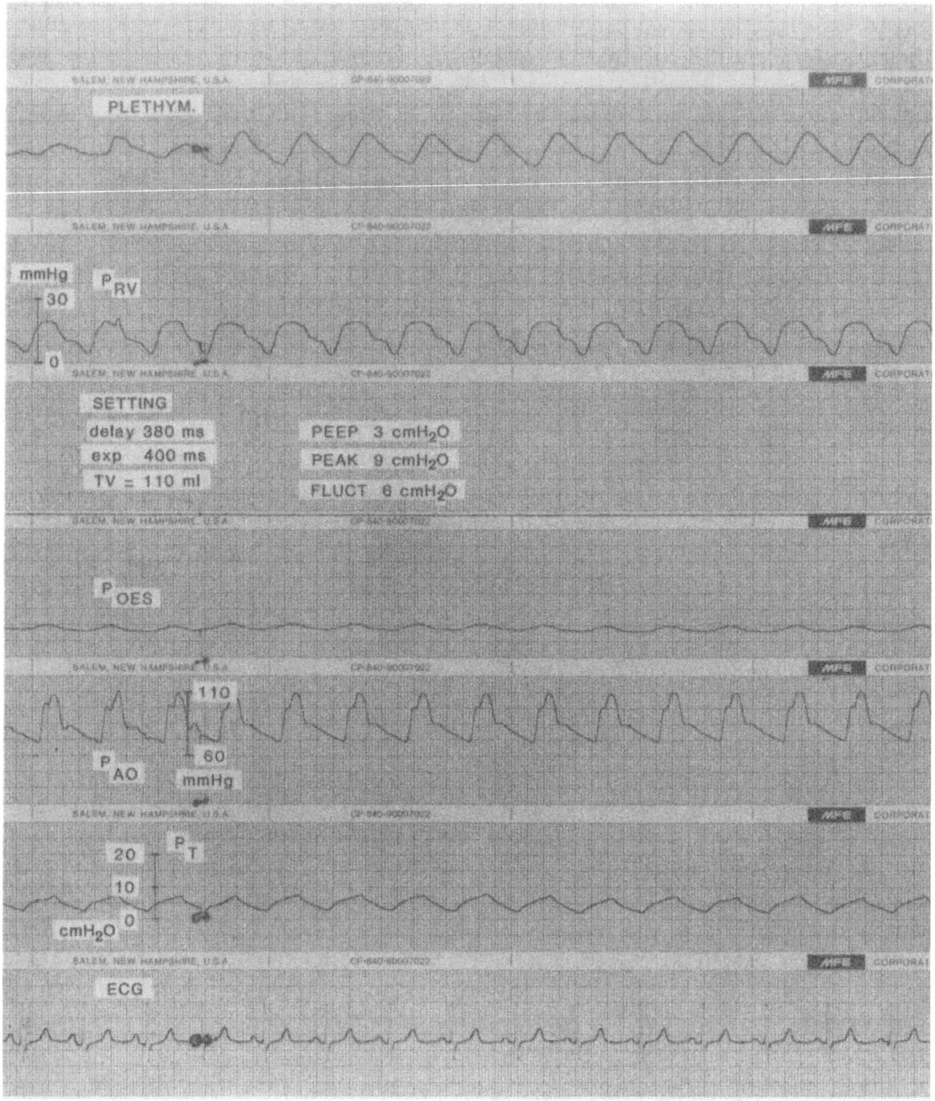

Fig. 5. Measurements as in Fig. 4, but lasting over 2 h

Table 1. Calculations based on micro-Astrup method

	pCO_2 (mmHg)	pO_2 (mmHg)	O_2 satuation (%)
IPPV	37.2	125	98
10 min jet	39.0	139	98
30 min jet	40.5	169	99
120 min jet	39.3	155	99

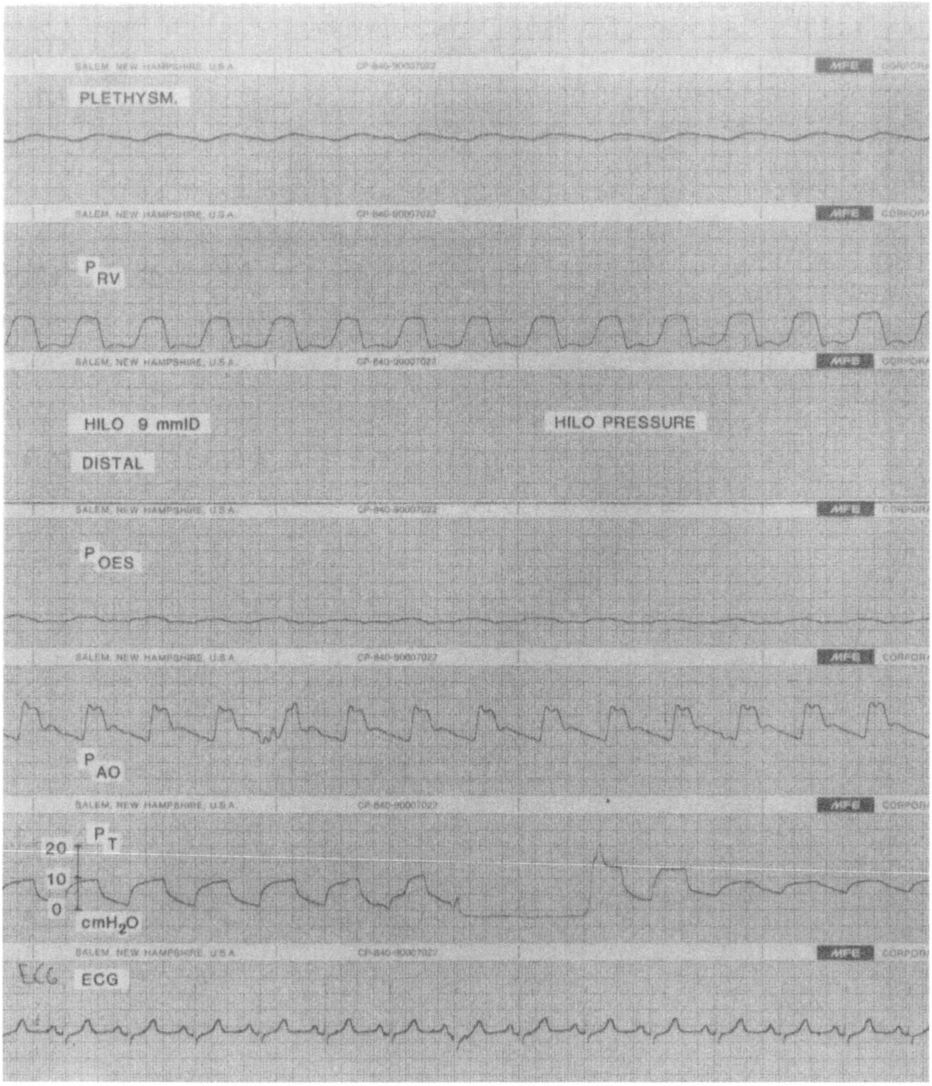

Fig. 6. Measurements as in Fig. 5. Intratracheal pressure was measured through the pressure opening of the Mallinckrodt Hi-Lo tube and via the tip of the Swan-Ganz catheter

We consider our way of heart-synchronized artificial ventilation to offer the highest efficiency of all available methods. It could prove very beneficial with intensive care patients and in those surgical procedures in which any influence on the circulation should be avoided. Asynchronous jet ventilation interferes in an unpredictable way with the pulmonary circulation. We feel that this should be avoided in clinical treatment. However, if asynchronous jet techniques are used, at least the intra-oesaphageal pressure and pulmonary artery pressure should be monitored. The Mallinckrodt Hi-Lo tube provides no real advantages for jet ventilation alone. It might be interesting using this tube for the new Bird IPPV jet ventilator. Pressure measurement through the special pressure lumen is unreliable and provides pressure fluctuations that are too low because of the design of the tube.

References

1. Dunnen WL den, Mostert T (1984) Pneumatic controlled circulation. In: Droh R, Erdmann W, Spintge R (eds) Anaesthesia – innovations in management. Springer, Berlin Heidelberg New York, pp 163–176
2. Dunnen WL den, Mostert T (1983) Pneumatic controlled circulation: PCC. In: Scheck PA, Sjöstrand UH, Smith RB (eds) Perspectives in high frequency ventilation, Elsevier, Rotterdam pp, 122–131

Innovative Aspects in Circulation

Physics of the Circulatory System

T. MOSTERT and W. L. DEN DUNNEN

Introduction

The physical properties and regulation of the circulation in over 40 patients and ten animals were studied and analysed. It was found that the circulation behaves exactly as a physical oscillator under all conditions. Beyond the neuronal effects on cardiac action, no trace could be found of another external influence upon the heart action. It therefore seems that regulation of the circulation is based solely on this phenomenon.

Methods

Blood pressure was measured intra-arterially in different arteries. The measurements were digitized and stored numerically on disk for later processing. It was found that measurements too far distal (e.g. radialartery) were too influenced by reflection against the capillaries, and we therefore measured blood pressure in the brachial or the femoralartery. Resolution of measurements was 0.05 mmHg, integrated over intervals of 20 ms. No electronic filtres were used to avoid phase shifting and distortion of the curves.

Discussion

Any pulsating driving force applied to an elastic medium generates a sinusoidal pressure wave of the form:

$$4P = A \sin (wt) \tag{1}$$

where P is pressure, A is amplitude of pulse pressure, w is cycle frequency and t is time. Thus, the arterial pressure wave due to cardiac ejection causes a sine wave (Fig. 1a). This sine wave is superposed on the initial pressure (P_0) at the moment of the opening of the aorta valve so that actual pressure during cardiac ejection becomes:

$$P = P_0 + A \sin (wt) \tag{2}$$

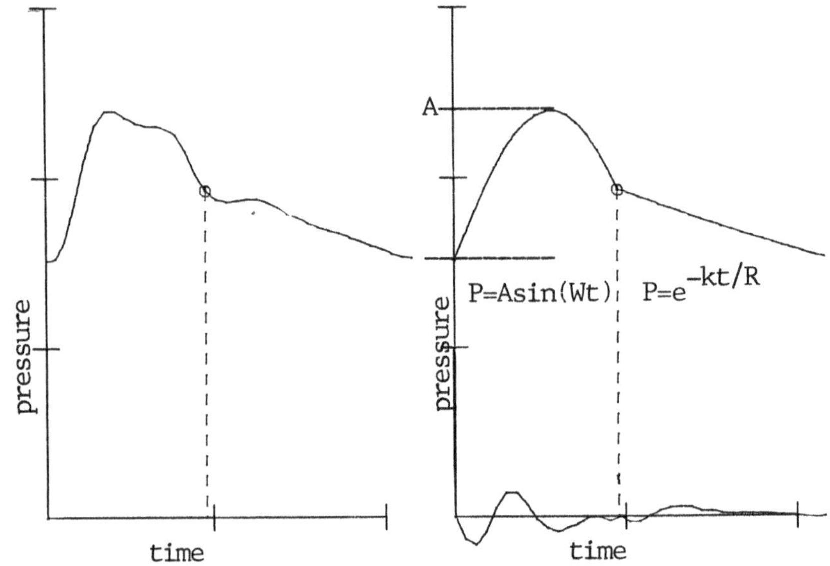

Fig. 1. Composition of an arterial pressure wave. *Left,* measured curve; *right,* calculated curve; *rigth, lower curve,* aortic valve circulation

The aorta valve closes when the differential pressure of the ventricle and the aorta equals the reactive force of the aortic valve. Any elastic system with or interacting with a mass has a natural, or resonance, frequency – similar to a spring and a mass, the pendulum of a clock, or the balance wheel of a watch. When brought into movement, it oscillates at a specific frequency. The cycle frequency of this oscillation *(w)* is given by the square root of the ratio of the force constant *(k)* and the mass *(m)*:

$$w = \sqrt{k/m} \tag{3}$$

This natural, or resonance, frequency is that frequency at which the dissipation of energy is at its minimum. It is in the nature of living organisms always to tune to the lowest dissipation of energy. Inducing a forced wave, lower or higher than the natural frequency, causes increased dissipation of energy. Because the heart is also elastic, ejection into the vascular system is near the resonance frequency of the system. As long as the driving force is near the natural frequency, the system resonates at this frequency. Hence, we are able to derive the cycle frequency or the time period of the driving force from the measured ejection curve and also the amplitude *(A)* or pulse pressure of this wave. The aortic valve prevents backflow of blood into the ventricle, so it acts like a rectifier diode in an electric circuit: flow is in only one direction. If the blood vessels were rigid, there would merely be a resistance, and the pressure curve would consist only of the positive part of the sine wave, or a single-rectified sine wave. But we are dealing here with a compliant system, obeying like a physical spring, obeying Hooke's law within a limited range.

We define the reactive force *(F)* as the product of the ejected volume and the force constant *(k)*. The force constant is the reciprocal of the vascular compliance.

$$F = Vk \tag{4}$$

Although the compliance of the blood vessels is not linear, Eq. 4 is satisfied within certain limits. During diastole, when the aorta valve is closed, the emptying curve of the arteries is always logarithmic and is given by (see also Fig. 1b)

$$P = P_0 \, e^{-tk/R} \tag{5}$$

We can calculate the ratio of the resistance *(R)* and the force constant *(k)* from Eq. 5:

$$\frac{k}{R} = \frac{-\ln P/P_0}{t} \tag{6}$$

If we substitute the force constant *(k)* for the reciprocal of compliance *(C)* in Eq. 6, we obtain:

$$RC = \frac{-t}{\ln P/P_0} \tag{7}$$

The *RC* factor is quite informative about the systemic circulation. In our opinion, this factor may prove very convenient in diagnosis and in patient monitoring.

Having defined the basic curve, there is still another component. When the aorta valve opens and closes, it generates an oscillation superposed on the basic curve (Fig. 1b). This wave, caused by the elastic properties of the aortic valve and its mass, has the natural frequency of the aortic valve. The principle is also the same as that in a physical spring. The amplitude is related to the velocity of the blood flow during the opening and closing of the valve. Because this wave is superposed, its integral value is always near zero. This is the case with the curve of every heart beat. Over 20000 heart cycles have been analysed, and each obeyed this principle.

There is only one temporary factor caused by the neuronal regulation for the association process in the brain. In the case of control of the circulation, we established that this is based on a simple harmonic oscillator in resonance. Figure 2 shows the interaction between respiration and the systemic circulation. The pressure wave during an inspiration is plotted over a pressure wave during the expiratory phase. During the negative intrathoracic pressure, part of the ejected blood from the right ventricle remains in the lungs so that the filling of the left ventricle is less, resulting in lower arterial pressure. The lungs are emptied more during expiration than during inspiration.

The total vascular impedance is slightly lower during the inspiratory phase, so the *RC* factor decreases, and the time period becomes shorter. We see in

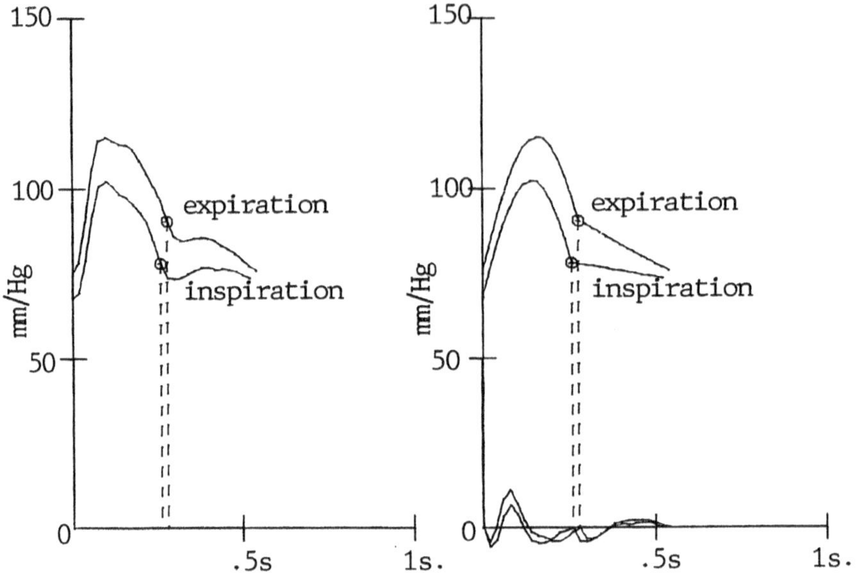

Fig. 2. Relationship between respiration and circulation. *Left,* measured curve; *right,* calculated curve; *right, lower curve,* aortic valve circulation

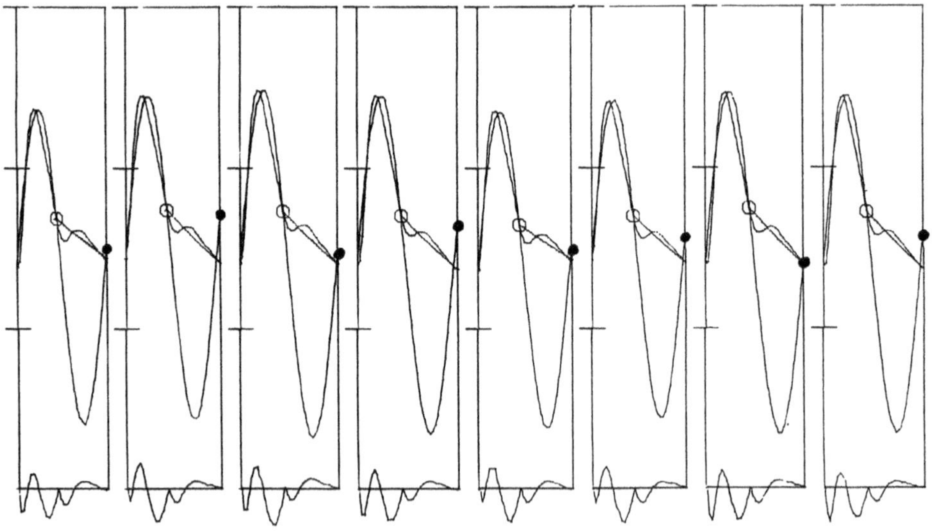

Fig. 3. Cycles of vascular impedance

Fig. 3 that for this reason, respiration causes modulation of the starting point of the next cardiac cycle respective to the cycle frequency around a node of the sine wave. Because the next cardiac cycle starts at a node, it is said to be in

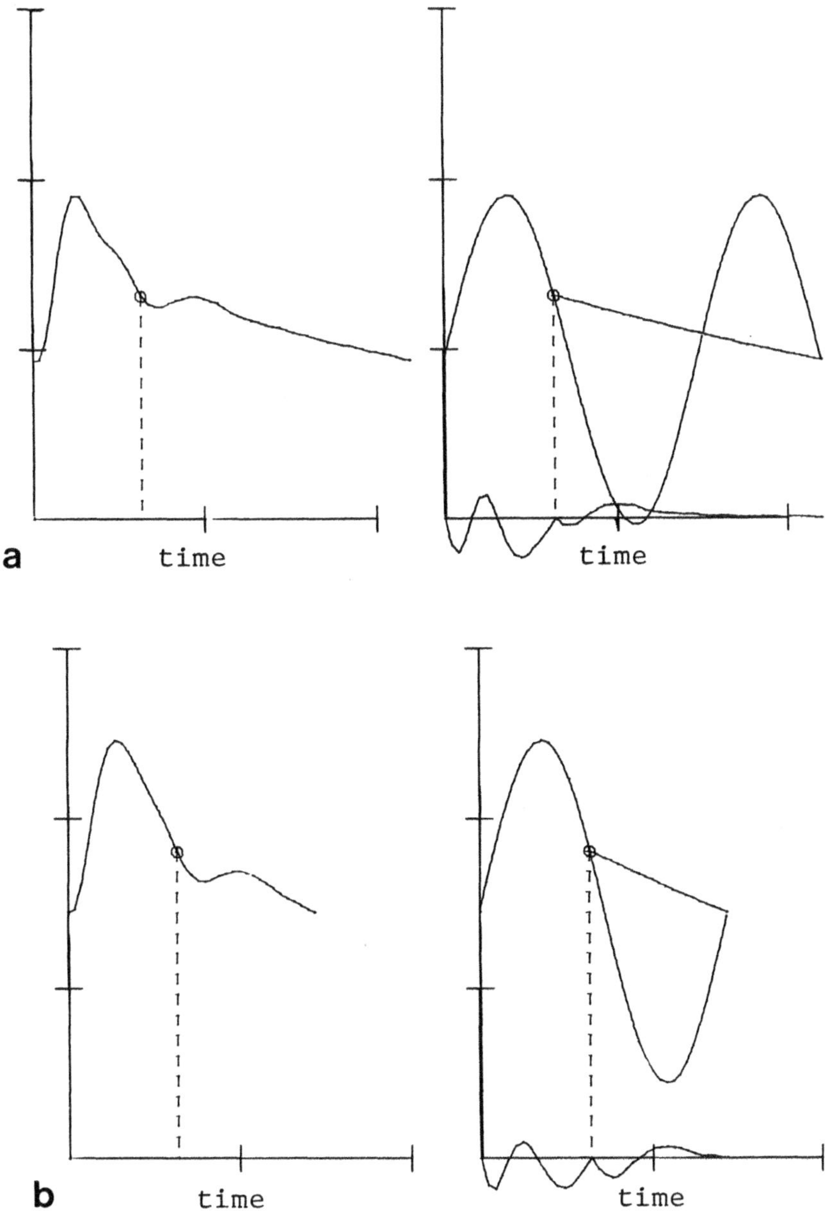

Fig. 4. Arterial pressure in stable relaxed phase **(a)** and in alert state **(b)**. *Left,* measured curve; *right,* calculated curve. The plotted sinus curve is extrapolated to show nodal points

resonance. The recording in Fig. 4 was made during a relaxed phase; it can be seen that the next cardiac cycle jumps to the next node. Although any differentiated starting moment of cardiac ejection increases energy dissipation, it is in a relatively highly efficient status.

In the deep sleep phase, energy dissipation is minimized by synchronizing respiration to the cardiac cycle. Nearly all circulatory and vascular parameters can be derived from a single measurement of arterial pressure and can be monitored every heart beat. Another interesting point is that all perfused parts of the body develop with the same resistance/compliance ratio, with the exception of those encapsulated in a rigid bone shell. For this reason, transplant organs must have the same resistance/compliance ratio as the receiving body otherwise there is poor tuning of the frequency, and a hypo- or hyperperfusion results, leading to disintegration or hypertrophy.

Conclusions

The circulatory system of the blood is a pure physical harmonic oscillator. The control of perfusion and circulation is performed only by the load and preload properties of the heart. Hence, it cannot be confirmed whether there exists any direct influence on heart action to regulate the circulation. Even the release of adrenaline into the heart serves the demand to follow the simple harmonic oscillator. The respiratory system varies the filling of the left ventricle and thus the ejection volume, resulting in variation of the arterial pressure.

Animal life is a composition, or matrix, of resonating structures. Each part of the body matches the natural frequency. Any differences of the natural frequencies decreases the half-life of that part as well as its immunological resistance. A cell structure in full resonance with the circulation has the highest immunological resistance.

The Effect of Peripheral Vascular Resistance on the Systolic Blood Pressure Difference Along an Extremity

J. C. DORLAS and A.W. CUIPER

Introduction

Two years ago we presented our studies on photoelectric plethysmography [1]. One of the conclusions was that the amplitude of the finger plethysmogram, in contrast to that of the ear, is very sensitive to changes in smooth-muscle tone of the arteriolar walls. The main reason for the strong reactivity of the finger plethysmogram is that the vascular walls in the finger are only α-sympathetically innervated [4]. We also presented our experience with a method to measure blood pressure in the finger continuously and yet non-invasively [2] with the new Finapres of Wesseling et al. [7]. We showed that during peripheral vasoconstriction systolic blood pressure in the fingers rise more than that in the upper arm, whereas it decreases more during peripheral vasodilation. It was also shown that systolic blood pressure in the radial artery reacted like that in the finger.

Most textbooks mention that pulse pressure increases along the arterial tree, and that this amplification effect is almost entirely due to a higher systolic pressure in the periphery. A few investigators also mention that this amplification effect is enhanced by peripheral vasoconstriction and diminished by peripheral vasodilation [6]. However, quantitative data are seldom given, and if they are, the results agree poorly with each other, probably because the resonance quality of the vascular bed is not known during these studies [3, 5]. Yet quantification is of great importance, since blood pressure is clinically measured either in the brachial artery or invasively in the radial artery, and changes in the amplification effect may seriously influence the interpretation of systolic pressures measured in this way.

Therefore, we have performed experiments in which various degrees of vasoconstriction and dilation are evoked in the hindleg of dogs. Blood pressure was measured in a proximal and a distal artery, and the differences in pressure between these two sites were correlated with peripheral vascular resistance and with the amplitude of a peripheral pulse plethysmogram. The latter is taken because, as mentioned above, we consider changes in its amplitude to give a good indication of changes in peripheral vasomotor tone. The preliminary results are reported in this paper.

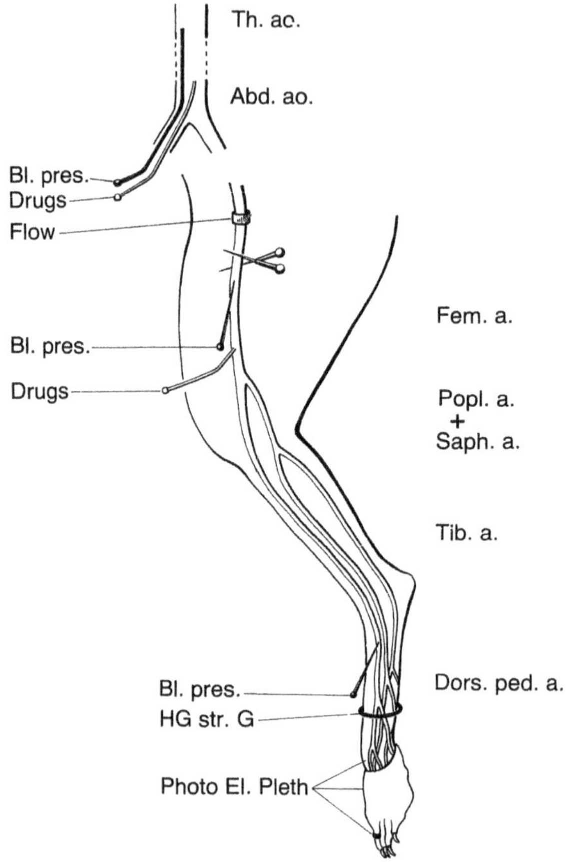

Fig. 1. Experimental set-up (see text)

Methods

Eleven experiments were performed in anaesthetized mongrel dogs of 20–25 kg (Fig. 1). Blood pressure was measured in the femoral artery and in the dorsal artery of the foot with the transducer connected directly to an intravascular cannula of 0.8 mm ID. Blood flow was measured electromagnetically in the femoral artery, comfortably proximal to the pressure cannula. From mean femoral blood pressure, and flow, peripheral vascular resistance was calculated (mmHg/ml, min). Furthermore, a photoelectric plethysmogram was recorded from an interdigital skin fold. From the contralateral groin a catheter tip manometer and an injection cannula were advanced up to the thoracic aorta and the abdominal aorta, respectively.

Small doses of noradrenaline (NA) were injected into the abdominal aorta, both with and without a continuous infusion of acetylcholine (AC) applicated with a pump via a side branch of the femoral artery. The doses of both drugs were varied, but we took care that the blood pressure in the thoracic aorta changed as little as possible. Figure 2 shows an example of an injection of 25 μg NA. Data were taken each time shortly before the injection and at he moment

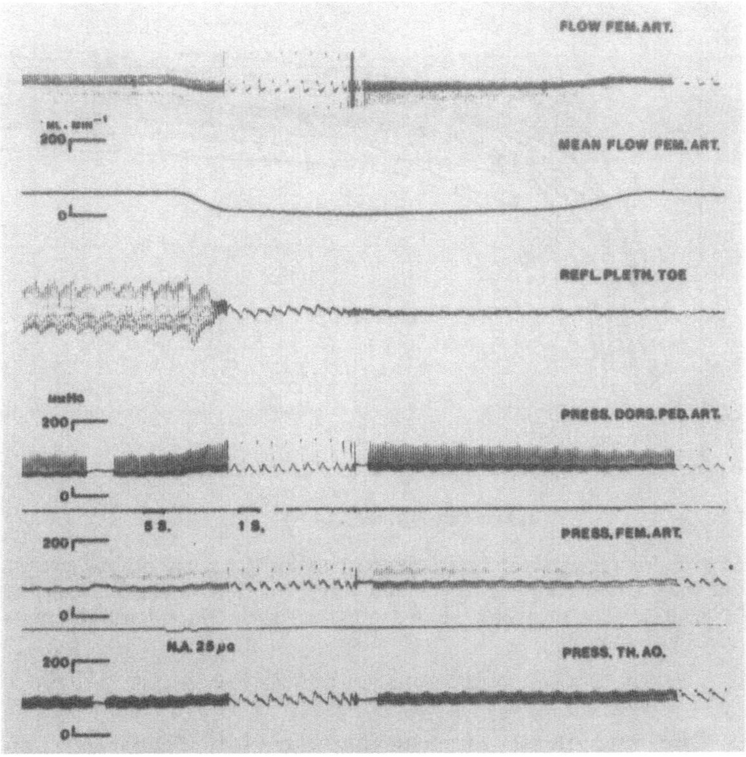

Fig. 2. Original recording of the effects of 25 μg noradrenaline *(N.A.)* injected in the abdominal aorta. Vertical arrow, Injection of N.A. followed by 1 ml saline. Interruptions in the pressure curves indicate mean pressure measurements

of maximum effect. For the calculations the differences in systolic, diastolic and mean pressure between the dorsal artery of the foot and femoral artery were correlated with peripheral vascular resistance as well as with the amplitude of the plethysmogram. The amplitude is given as percentage of the maximum amplitude found in the same dog.

Results

Figure 3 shows the relationship between the systolic pressure difference (ΔP_S; dorsal artery of the foot minus femoral artery) and the peripheral vascular resistance *(R)* for all dogs separately. The curves are pooled to give an average blood pressure level. It shows that in the low-resistance area ($R < \pm 1$) a small change in R corresponds to a large change in ΔP_S, while the relationship becomes much more gradual in the high-resistance area ($R > \pm 1$). Although the range is rather wide, the trend is obvious. The thick line, representing the mean curve, shows that the sharpest bend occurs at about 10 mmHg.

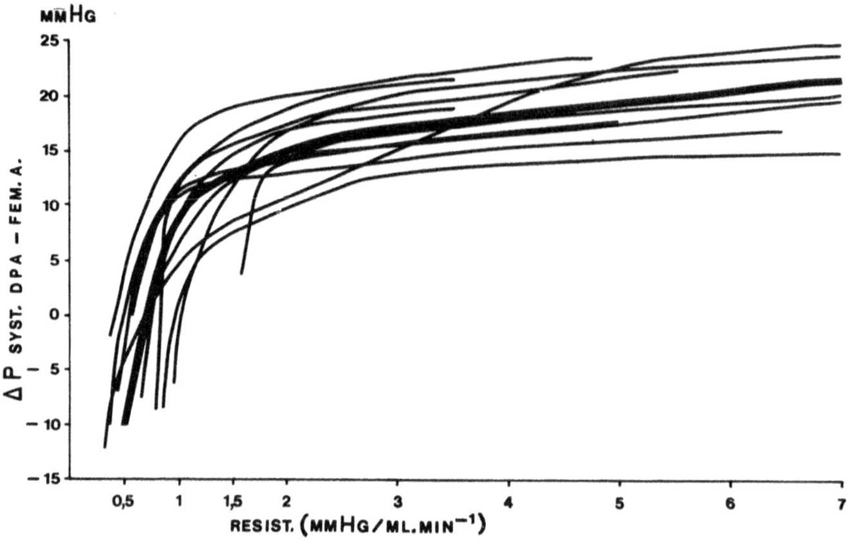

Fig. 3. Diagram of the best fitting curves for the relationship between ΔP_S (dorsal artery of the foot, *d.p.a.*, minus femoral artery) and R in the 22 experiments separately, but pooled to give an average blood pressure level. *Thick line* represents mean curve for all dogs collectively

The data of all dogs together are plotted semilogarithmically in Fig. 4. The dots show the relationship between the dorsal artery of the foot and the femoral artery. Although the range is still rather wide, the best fitting line proved to be slightly curved and not straight. Therefore, a best fitting broken line was calculated. This shows that, despite the semilogarithmic plot, the slope in the low-resistance area is still steeper than that in the high-resistance area. The break point is found at about 10 mmHg, corresponding to an R value of about 1. For comparison, the relationship with systolic pressure in the thoracic aorta is also given, showing that in the high-resistance area the systolic pressure difference between the dorsal artery of the foot and the thoracic artery is even greater than that between the dorsal artery of the foot and the femoral artery. The two lower curves of Fig. 4, representing the relationship, between R and the diastolic and mean pressure differences between the dorsal artery of the foot and the femoral artery, show about the same breaking point but a much weaker slope.

Figure 5 shows the relationship between the systolic pressure difference (between dorsal artery of the foot and femoral artery) and the plethysmographic amplitude. Because the pressure data are pooled to give an average level, the lines meet at about 10 mmHg, which corresponds to an amplitude of 30%–40%. Furthermore, the separate curves show that during vasoconstriction from 15% to 5% the systolic pressure in the dorsal artery of the foot becomes much higher than that in the femoral artery, whereas during vasodilation from 40%–100% ΔP_S decreases much more gradually.

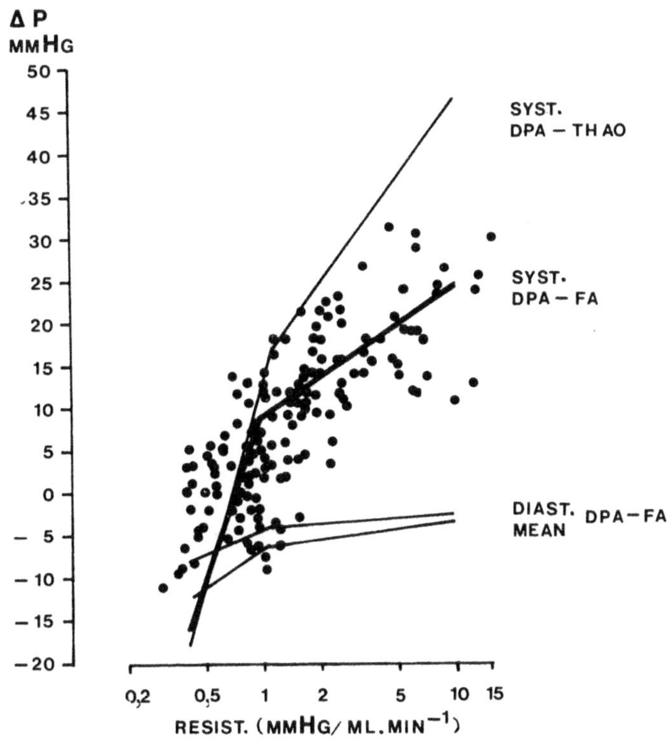

Fig. 4. Semilogarithmic graph of the relationship between Δ P_S and R for all dogs collectively. *Dots* and *thick broken line* represent the relationship for Δ P_S between dorsal artery of the foot *(DPA)* and femoral artery *(FA)* and *upper line* that for Δ P_S between DPA and thoracic aorta *(THAO)*. *The two lower broken lines* show the relationship for the diastolic and mean pressure differences between *DPA* and FA

Discussion

Our results show a direct relationship between the degree of peripheral vascular resistance and the difference in arterial blood pressure along an extremity. The effect is by far the strongest for systolic pressure, but in the low-resistance area the effect on diastolic and mean pressure can also not be neglected. When R is plotted semilogarithmically, a break point is found at $R = \pm 1$, corresponding with a Δ P_S of + 10 mmHg. We assume that these data represent normal resting conditions. From this point onwards an increase in R results in a rise in systolic pressure in the dorsal artery of the foot which on average becomes 25 mmHg higher than in the femoral artery and 45 mmHg higher than in the thoracic aorta at an R value of 10.

Clinically this result means that systolic blood pressure measured in a peripheral artery may be very misleading. This is the more so because, for instance, during anaesthesia, data on vascular resistance are not available. The only practical approximation is offered by peripheral pulse plethysmography, because changes in its amplitude reflect the occurrence of vasoconstriction or

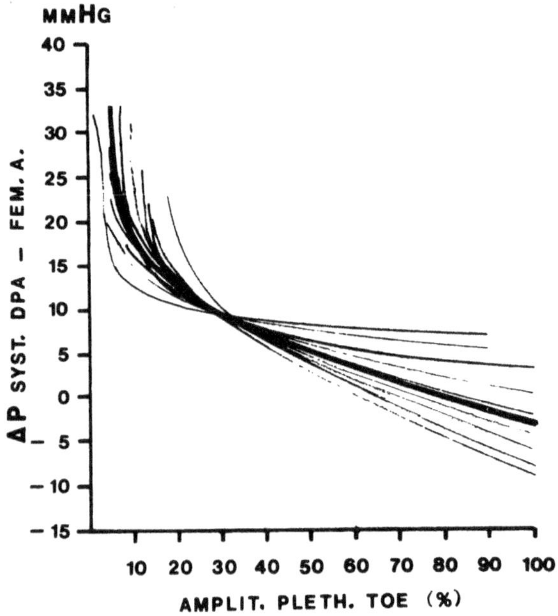

Fig. 5. Graphic relationship between ΔP_S (dorsal artery of the foot, *DPA,* minus femoral artery, *(fem. a.))* and the amplitude of the peripheral pulse plethysmogram for all 11 dogs separately, but pooled to give an average blood pressure level. *Thick line* represents mean curve for all dogs collectively

vasodilation. Since a plethysmographic amplitude of roughly 30%–40% corresponds in our study to a ΔP_S of 10 mmHg and an R of $\pm$ 1, this amplitude may be assumed to be the average for normal resting conditions. A decrease in amplitude from this point initially corresponds to a gradual increase in the systolic pressure difference, but between 15% and 5% the rise in ΔP_S becomes very steep and unpredictable. Probably the plethysmogram is not sensitive enough at low amplitudes. On the other hand, an increase in amplitude from the resting value to 100% corresponds to a gradual and relatively small decrease in ΔP_S. Although this seems to be a clinically useful finding, the small decrease in ΔP_S is conflicting with the steep change in ΔP_S in the low-resistance area (Figs. 3, 4; $R < + 1$). This discrepancy may be partly attributed to the fundamentals of plethysmography. The amplitude is almost always determined by the smooth-muscle tone of the arteriolar walls but is determined by the intravascular pulse pressure in the dilated state [1]. However, this explanation cannot account for the steep fall in ΔP_S in the vasodilated state. Therefore, we are convinced that the discrepancy is at least partly caused by an opening up of arteriovenous anastomoses (AVAs). These anastomoses are abundant in the hands and feet, and when open they cause a marked fall in peripheral blood pressure in particular [2]. However the plethysmographic amplitude, being dependent only on the arterioles, shows no further increase because they must be widely dilated already. The effect of the shunts may even be so large that a steal effect occurs, leading to a decrease rather than an increase in amplitude.

In conclusion, the occurrence of marked peripheral vasoconstriction or vasodilation may lead to a real but misleadingly high or low systolic blood pressure in peripheral arteries. Changes in the amplitude of the peripheral pulse plethysmogram give an indication only as to the direction of the pressure deviation. Only in a small margin around the normal resting vascular tone (an amplitude between 15% and 50%) is there some correlation with the degree of systolic pressure deviation. Mean arterial blood pressure gives a much better reflection of the central situation, although during marked peripheral vasodilation it may give too-low values as well, possibly due to open AVAs.

References

1. Dorlas JC, Nijboer JA, Edema JA, Prins JOJ (1985) A new approach of photo-electric plethysmography. In: Droh R, Erdmann W, Spintge R (eds) Anaesthesia – innovations in management. Springer, Berlin Heidelberg New York
2. Dorlas JC, Nijboer JA, Butijn WT, Cuiper AW, Wesseling KH, van den Hoeven GMA, Settels JJ (1985) A continuous non-invasive bloodpressure measurement which is controlled by fingerplethysmography. In: Droh R, Erdmann W, Spintge R (eds) Anaesthesia – innovations in management. Springer, Berlin Heidelberg New York
3. Kroeker EJ, Wood EH (1955) Comparison of simultaneously recorded central and peripheral arterial pressure pulses during rest, exercise and tilted position in men. Circ Res 3:623–631
4. Nijboer JA, Dorlas JC (1985) Comparison of plethysmograms taken from finger and pinna during anaesthesia. Br J Anaesth 57:531–534
5. O'Rourke M, Avolio AP (1980) Pulsatile flow and pressure in human systemic arteries. Circ Res 46:363–372
6. O'Rourke M, Taylor MG (1966) Vascular impedance in the femoral bed. Circ Res 18:126–139
7. Wesseling KH, de Wit B, Settels JJ, Klawer WH (1982) On the indirect registration of finger bloodpressure after Penaz. Funkt Biol Med 1:245–250

Haemodilution and Myocardial Ischaemia – Studies with Fluorocarbons and Dextran in Pigs

N. S. FAITHFULL, M. FENNEMA and W. ERDMANN

Myocardial ischaemia in surgical patients is an important risk factor, particularly if the patient has suffered recent myocardial infarction; cardiac reserves are compromised then, and tissue oxygen flux may be limited. Reduction of blood viscosity by haemodilution may allow output to increase without increase in cardiac work, but due to the inability of most dilutants to carry oxygen, oxygen transport does not increase and may be decreased.

Fluorocarbon-containing plasma substitutes in the form of Fluosol-DA 20% (FDA) have been released for limited clinical trials and can carry 1 ml per 100 ml emulsion per 100 mmHg PO_2 (Naito and Yokoyama 1978). These figures have, however, been challenged, and it appears that the true O_2 transport of FDA is near to 0.75 ml per 100 mmHg (Zander and Makowski 1982; Grote et al. 1985). This amount is small, but it is nevertheless 2.5 times that of plasma. In order to transport significant amounts of O_2 in the arterial blood, the inspired concentration of O_2 is usually raised, and under these conditions, due to the linear relationship between blood PO_2 and O_2 content, the extraction of available oxygen is very high.

In circumstances where blood lost during severe haemorrhage cannot be replaced by transfusion, for instance, in the absence of transfusion facilities or of suitably cross-matched blood or if blood is refused on religious grounds, replacement of lost blood with FDA should offer definite advantages over the use of conventional plasma substitutes, particularly where the cardiac reserve is limited. The present paper addresses this question, using a myocardial ischaemia model in pigs.

Methods

Three groups of five juvenile Yorkshire pigs (body weight 23.5–27 kg) were anaesthetized with an intraperitoneal induction dose of thiopentone 30 mg/kg. The trachea was intubated, and the lungs were ventilated with 66% nitrous oxide in oxygen and 1.0%–1.5% halothane. Muscular relaxation was obtained using a continuous infusion (0.5 mg/kg per hour) of pancuronium bromide. Systemic arterial pressure was monitored via the left femoral artery, and pulmonary artery pressure, pulmonary capillary wedge pressure and cardiac output were obtained using a 7-F gauge thermodilution Swan-Ganz catheter inserted via the left femoral vein. Central venous pressure was monitored via a catheter in the right femoral artery. All the above pressures were transduced and displayed on a Grass 7D polygraph and ink-writing oscillograph.

Arterial and mixed venous blood gas and acid-base state, haemoglobin concentration and oxyhaemoglobin saturation were regularly estimated using a Radiometer ABL1 acid/base laboratory and a Radiometer OSM1 cooximeter. When FDA had been administered, readings from the cooximeter were inaccurate, and oxyhaemoglobin saturation was calculated from the computer subroutine described by Kelman (1966).

A thoracotomy was performed, and the pericardium was opened. In order to assess the effects of haemodilution on the myocardial oxygen tension (PmO_2) in both the ischaemic and non-ischaemic areas, four steel-protected gold microelectrodes 200 μm in diameter were inserted approximately 3 mm into the myocardium of the lower anterior wall of the left ventricle. These were placed so that two were in and two were outside the infarct area.

The left anterior descending coronary artery (LAD) was occluded at the junction of its lower and middle thirds, and 1 h later two of the groups of animals were bled (20 ml/kg body weight) and the volume deficit was replaced in one group by FDA and in the other by 5% dextran solution (mol. wt. 40000). The third (control) group received no treatment. The experiments were terminated after 5 h of ischaemia by ventilating the animals with 100% nitrous oxide. In order to take advantage of the oxygen-carrying capacity of the fluorocarbons, the animals were ventilated with 0.5% halothane in 100% oxygen for at least 0.5 h prior to LAD occlusion.

Cardiovascular and oxygenation variables were calculated using a Phillips P2000 microcomputer, and statistical analysis was performed with an Apple IIe microcomputer. Statistical significance was assessed using the paired or unpaired Student's t tests, and the null hypothesis was rejected at a p value of less than 0.05.

Results

Haemodilution caused mean increases in cardiac output in both the dextran and the FDA-treated animals. Although the increases were not statistically significant, the FDA group did have substantially greater cardiac outputs immediately after dilution than did the control group. At this time the mean values for whole body oxygen delivery (QO_2) were also higher than in the controls (Fig. 1), and at no time up to 4 h after blood volume replacement was QO_2 significantly decreased below pre-dilution values. Significant decreases were seen in QO_2 in the dextran groups at all times post-dilution. Oxygen extraction coefficients did not differ between FDA and control groups, but at 3 and 4 h post-dilution those of the dextran group were significantly higher than those of the FDA group.

Left ventricular work (LVW) was unaltered in all groups (Fig. 2), but QO_2 per unit LVW was significantly decreased following haemodilution with dex-

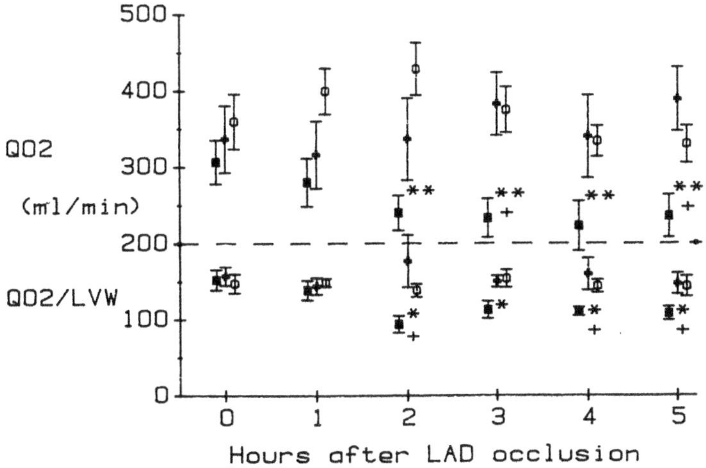

Fig. 1. Changes in whole body oxygen flux *(QO_2)* and flux per kilogram meter of left ventricular minute work *(QO_2/LVW)* following myocardial infarction in pigs. Haemodilution with dextran or Fluosol-DA 20% was performed between hours 1 and 2. Dextran group indicated by *closed squares,* Fluosol group by *open circles* and control group by *crosses.* Significant difference from pre-infarct value (0 h) indicated by *asterisks* ($p < 0.05$; $p < 0.01$) and difference from the control group by *crosses* (+ $p < 0.05$)

tran and was less than in both other groups (Fig. 1). Mean systemic arterial pressure (Fig. 3) tended to decrease following haemodilution and at 1 h after completion of haemodilution was significantly lower in both the FDA and dextran groups than in the control group.

Considerable changes in the haemodynamics of the pulmonary circulation were to be seen following haemodilution with FDA. Immediate and highly significant increases in pulmonary arterial pressures (Fig. 3) resulted in large, significant increases in right ventricular work (Fig. 2), and mean minute work values increased by 140%.

Resting PmO_2 levels were very variable. As previosly noted (Faithfull et al. 1985a), the pattern of PmO_2 changes following occlusion varied from animal to animal, and it was almost impossible to predict the direction of change that would occur in an individual electrode. The electrode showing the greatest percentage decrease in PmO_2 over the 1st h of occlusion was taken as being in the most ischaemic area (Faithfull et al. 1985b). Percentage changes in PmO_2 from this electrode are shown in Fig. 4. Significant decreases in oxygen tensions occurred 1 h after LAD occlusion, and in the control and dextran groups PmO_2 slowly decreased thereafter.

At 2 h after LAD occlusion, a point coinciding with completion of haemodilution, PmO_2 was no longer significantly lower than before occlusion in the FDA-diluted group. The oxygen tensions in this group showed a wide range of values, and the mean values rose slowly until a sudden and sustained rise began 1 h after completion of haemodilution. At the end of the experiments (3 h after haemodilution) mean values of PmO_2 had reached 92.9% of the pre-occlusion values.

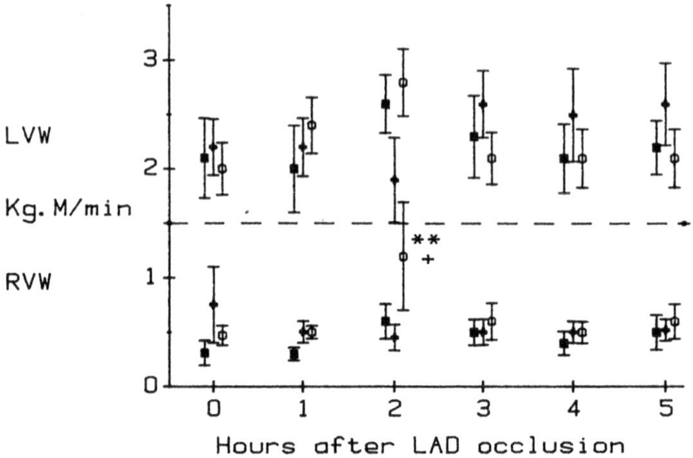

Fig. 2. Changes in left ventricular minute work *(LVW)* and right ventricular minute work *(RVW)* following myocardial infarction in pigs. Haemodilution with dextran or Fluosol-DA 20% was performed between hours 1 and 2. Dextran group indicated by *closed squares,* Fluosol group by *open circles* and control group by *crosses.* Significant difference from pre-infarct value (0 h) indicated by *asterisks* (* $p < 0.05$; ** $p < 0.01$) and difference from the control group by *crosses* (+ $p < 0.05$)

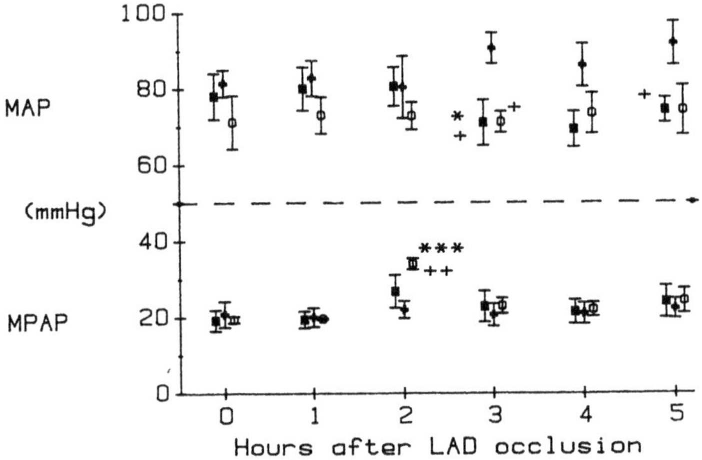

Fig. 3. Changes in mean systemic arterial pressure *(MAP)* and mean pulmonary arterial pressure *(PAP)* following myocardial infarction in pigs. Haemodilution with dextran or Fluosol-DA 20% was performed between hours 1 and 2. Dextran group indicated by *closed squares,* Fluosol group by *open circles* and control group by *crosses.* Significant difference from pre-infarct value (0 h) indicated by *asterisks* ($p < 0.05$; *** $p < 0.001$) and difference from the control group by *crosses* (+ $p < 0.05$)

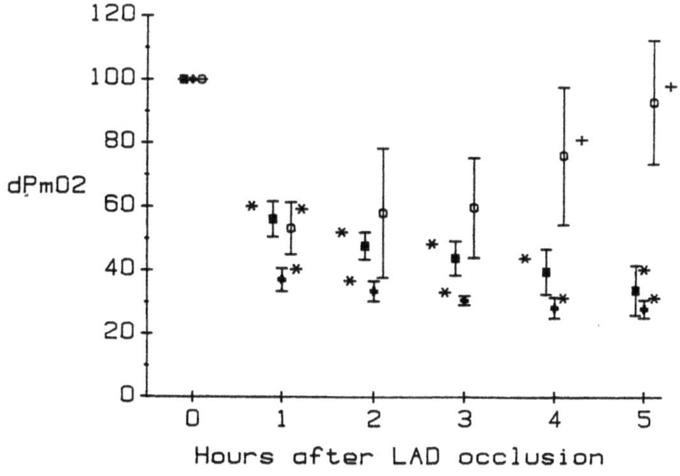

Fig. 4. Percentage changes in myocardial oxygen tension *(dPmO2)* in the most ischaemic area of a fresh myocardial infarct in pigs. Haemodilution with dextran or Fluosol-DA 20% was performed between hours 1 and 2. Dextran group indicated by *closed squares,* Fluosol group by *open circles* and control group by *crosses.* Significant difference from pre-infract value (0 h) indicated by *asterisks* (* $p < 0.05$) and difference from the control group by *crosses* (+ $p < 0.05$)

Discussion

The above results demonstrate a striking difference between cardiovascular responses to haemodilution with FDA in the systemic and pulmonary circulations. The pulmonary hypertensive response would appear to be species-dependent, and we have observed that a dose of 0.5 ml may cause a fatal reaction in pigs. These reactions, which were not prevented by prior administration of large doses (3 g) of hydrocortisone, were characterised by massive increase in pulmonary artery pressures and extreme systemic hypotension. Enormous dilatation of the right ventricle and acute tricuspid valve insufficiency were also seen. Death ensued within a matter of minutes of right heart failure. In some cases, spontaneous decreases in the pumonary vascular resistance occurred after about 5–10 min, with consequent restoration of systemic circulation. Ventricular fibrillation often occurred then, presumably as a result of massive release of acid metabolites into the circulation. Restoration of the circulation was often so sudden that this occurred in spite of the administration of large amounts of sodium bicarbonate.

There does not appear to have been a great deal of published work on the use of perfluorochemicals (PFCs) in pigs. Grunert et al. (1982) studied biochemical changes and organ uptake of PFCs following 20% blood volume replacement with FDA but did not report any reactions. They did, however, mention "increasing pulmonary deterioration following substitution". Engleman et al. (1983) used FDA as a priming agent during cardiopulmonary bypass in pigs and noted that while systemic pressures remained stable during bypass, significant hypotension occurred during recovery. This was accompanied by greatly increased pulmonary arterial pressures and a two- or threefold increase in pulmonary vascular resistance. They reported subsequent microscopy findings of an "interstitial infiltrate ... and apparent loss of pulmonary vasculature". These latter findings agree with our own microscopy findings, which suggest the occurrence of massive pulmonary vasospasm in response to FDA.

The reactions to fluorocarbon emulsions in pigs described above are in direct contrast to those in dogs, which almost always develop acute hypotension following intravenous injection of fluorocarbons. Many reports of these reactions have appeared, although many authors have considered them to be of little importance; Clark et al. (1970) commented that further infusion of fluorocarbon emulsions would promptly overcome the hypotension. Maki et al. (1972) attributed canine hypotensive reactions to anaphylactoid shock, while Ohyanagi and Mitsuno (1975) considered the reaction to be a result of serotonin and histamine release. Following extensive investigations, Suyama et al. (1979) concluded that the hypotension was due to direct and indirect effects of FDA acting through an α-adrenergic blocking action. This explanation cannot, however, account for the massive increases in systemic peripheral resistance seen during the hypotensive episode (N. S. Faithfull and S. M. Cain, unpublished). Even though the acute haemodynamic reactions have usually completely recovered after about 1 h, there is evidence that inhomogeneity of microcirculatory perfusion can persist for some hours after the reactions (Faithfull et al. 1986).

In humans the few cardiovascular reactions that have occurred have been more similar to those in pigs than those to in dogs. The first reports came from Tremper et al. (1981), and these were fully analysed in Tremper et al. (1984). In the reactions reported, increases of pulmonary arterial pressure were observed leading in some instances to moderate decreases in cardiac output. In one patient they were able to blunt the reactions, which subsided spontaneously within 3 min, by massive doses of methylprednisolone. These reactions were associated with similarly transient asymptomatic decreases in white blood cell count. Waxman et al. (1984) reported a reaction involving systemic hypotension and no increase in pulmonary arterial pressures; the cardiac output remained stable. These changes occurred in spite of prophylactic administration of 2 g methylprednisolone.

Vercellotti et al. (1982) have established that FDA can activate complement in human plasma in vitro via the alternative pathway. They further found that Pluronic F68, one of the emulsifiers present in FDA, was the component responsible for this effect. This complement activation caused aggregation of granulocytes, and the researchers were able to induce in vivo granulocytopenia in rabbits. Decreases in PaO_2 were also observed, and both these effects, which again were caused by Pluronic F68, could be significantly blunted by methylprednisolone.

While it is of considerable interest that almost all reactions have been reported in Caucasians, Ohyanagi et al. (1984) recently reported acute transient (15-min duration) leukopenia and thrombopenia in four out of seven (presumably Japanese) patients; no cardiovascular reactions were reported. Vercellotti et al. (1982) pointed out that lipoprotein profiles differ between

Americans and Japanese and have suggested that dietary factors may play a role. This might account for differences in complement activation or in the body's responses to it.

The responses of the systemic circulation to haemodilution with FDA were similar in the pigs employed in these experiments to those reported during therapeutic administration to human subjects. However, as no haemodilution studies have been performed in man, the responses are not strictly comparable. Studies on the administration of FDA during surgery using 2:1 nitrous oxide and oxygen mixtures (Nishimura et al. 1983) revealed nonsignificant increases in cardiac output with decreases in systemic vascular resistance 30 min after administration, accompanied by significant increases in whole body oxygen flux. However, in cases where the oxygen flux is inadequate for normal oxidative metabolism, the administration of FDA, while increasing oxygen flux, may result in the reduction of cardiac output as the oxygen delivery becomes adequate, and cardiac drive is decreased (Tremper et al. 1984).

In conclusion, these studies suggest that the use of FDA for replacement of blood loss during surgery in patients with recent myocardial infarction is acceptable. More importantly, it can result in improved oxygen delivery and improvement in the oxygenation status of the ischaemic myocardium.

Summary

A study was performed to investigate the effect of haemodilution with the oxygen-transporting plasma substitute on the ischaemic myocardium. Juvenile pigs (weight, 23.5–27 kg) were anaesthetised with intraperitoneal thiopentone (30 mg/kg) and maintained with 1%–1.5% halothane in N_2O and O_2 (2:1). The animals were ventilated under continuous pancuronium infusion (0.5 mg/kg per hour). Systemic arterial pressure and central venous pressure were monitored for pulmonary artery pressure, pulmonary capillary wedge pressure and cardiac output. A thoracotomy was performed and steel-protected gold microelectrodes 200 μm in diameter were inserted approximately 3 mm into the myocardium of the lower anterior wall of the left ventricle. After stabilisation under 0.5% halothane in oxygen (to ensure maximal oxygen transport by FDA when administered) the distal third of the LAD was occluded. After 1 h two groups of five animals were bled (20 ml/kg), and the blood lost was immediately replaced with similar volumes of FDA or 5% dextran. These procedures were complete by 2 h after LAD occlusion. A third group served as a no-treatment control. Immediately following haemodilution, FDA-treated animals had significantly greater cardiac outputs than did control animals, and the mean values for QO_2 were higher than in the controls (although not significantly so). At no time up to 4 h after blood volume replacement was QO_2 significantly decreased below pre-dilution values, whereas significant decreases were seen in the dextran group. Oxygen extraction coefficients were not different in the control and FDA groups but at 3 and 4 h post-dilution those of the dextran group were significantly lower than those of the FDA group. LVW was unaltered in any of the groups, but QO_2 per unit LVW was significantly smaller following haemodilution with dextran than it was in the other two groups. Pulmonary arterial pressures, wedge pressures and right ventricular work were all significantly increased immediately following blood replacement with FDA – a normal occurrence in pigs. Myocardial oxygen measurements indicated progressive improvement of oxygen tensions in the most hypoxic areas in the FDA group but progressive decrease in tensions in the other two groups.

References

Clark LC, Kaplan S, Becattini F (1970) The physiology of synthetic blood. J Thorac Cardiovasc Surg 60:757–773

Engleman RM, Dobbs WA, Rousou JH, Anisimowicz L (1983) Use of Fluosol-DA during open heart surgery. Prog Clin Biol Res 122:273–282

Faithfull NS, Erdmann W, Fennema M (1985a) Oxygen supply to the myocardium. Adv Exp Med Biol 180:411–423

Faithfull NS, Erdmann W, Fennema M (1985b) Oxygen tensions in the ischaemic myocardium following haedmodilution with fluorocarbons or dextran. Br J Anaesth 57:816P–817P

Faithfull NS, King CE, Cain SM (1986) Peripheral vascular responses to fluorocarbon administration. Microvasc Res 13,4

Grote J, Steuer K, Muller R, Sontgerath C, Zimmer K (1985) O_2 and CO_2 solubility of the fluorocarbon emulsion Fluosol-DA 20% and O_2 and CO_2 dissociation curves of bloodFluosol-DA 20% mixtures. Adv Exp Med Biol 191:453–461

Grunert A, Hirlinger WK, Herrmenn N, Petutschnigk D (1982) Experiences with Fluosol in animal experiments. In: Frey R, Beisbarth H, Stossek K (eds) Oxygen carrying colloidal blood substitutes. Zuckschwerdt, Munich, pp 86–90

Kelman GR (1966) Digital computer subroutine for the conversion of oxygen tension into saturation. J Appl Physiol 21:1375

Maki T, Hori M, Idezuki Y (1972) Severe circulatory derangement induced by intravenous fluorocarbon emulsion. J Surg Res 2:90–95

Naito R, Yokoyama K (1978) Perfluorochemical blood substitutes. Green Cross, Osaka (Technical information series, no 6)

Nishimura N, Hiranuma N, Sugi T (1983) Cardiovascular effects of various colloidal solutions during major abdominal surgery. Crit Care Med 11:940–942

Ohyanagi H, Mitsuno T (1975) Biophysiological effects of perfluorochemicals as artificial blood. In: Proceedings of the 10th International Congress of Nutrition – Symposium on PFC Artificial Blood. Igakushobo, Osaka, pp 21–54

Ohyanagi H, Nakaya S, Okamura S, Saitoh Y (1984) Surgical use of Fluosol-DA in Jehovah's Witness patients. Artif Organs 8:10–18

Suyama T, Watanabe M, Hanada S, Yano K, Yokoyama K, Naito R (1979) Pharmacological analysis of the mode of transient hypotensive action of Fluosol-DA found in the dog. In: Proceedings of the 4th International Symposium on PFC Blood Substitutes. Exerpta Medica, Amsterdam, pp 257–266

Tremper KK, Levine E, Friedman A, Shoemaker WC, Katz R (1981) The preoperative treatment of severely anemic patients with a perfluorochemical blood substitute, Fluosol-DA 29%. Anesthesiology 56:A9

Tremper KK, Vercellotti GM, Hammerschmidt DE (1984) Hemodynamic profile of adverse clinical reactions to Fluosol-DA 20%. Crit Care Med 12:428–431

Vercellotti GM, Hammerschmidt DE, Craddock PR, Jacob HS (1982) Activation of plasma complement by perfluorocarbon artificial blood: probable mechanism of adverse pulmonary reactions in treated patients and rationale for corticosteroid prophylaxis. Blood 59:1299–1304

Waxman K, Cheung K, Mason R (1984) Hypotensive reaction after infusion of a perfluorochemical emulsion. Crit Care Med 12:609–910

Zander R, Makowski HV (1982) Life without haemoglobin? In: Frey R, Beisbarth H, Stossek K (eds) Oxygen carrying colloidal blood substitutes. Zuckschwerdt, Munich, pp 133–141

Innovative Aspects in Different Medical Specialities

Sevoflurane – A New Inhalant Anesthetic Agent

T. MURAKAWA, Y. SATOH, A. MATSUKI, and T. OYAMA

Introduction

Sevoflurane, $(CF_3)_2CHOCH_2F$, is a new inhalation anesthetic produced by Travenol Laboratories in the United States. The name sevoflurane was derived from the presence of seven florides per molecule. The molecular weight, specific gravity, and boiling point of sevoflurane are similar to those of isoflurane, enflurane, and halothane. Its blood gas partition coefficient is lower than that of other inhalation anesthetics. The mean anaesthetic concentration is estimated to be approximately 2%. The anesthetic potency of sevoflurane is thus less than that of other agents. This agent has pleasant, ether-like odor. The biometabolic rate of sevoflurane is reported to be about 2%, which is as low as that of enflurane.

No study has been reported on the effects of sevoflurane anesthesia on endocrine function. Therefore, we undertook to investigate the effect of sevoflurane anesthesia and surgical stress on plasma levels of stress hormones and on hemodynamic changes during orthopedic surgery.

Methods

Twelve patients ranging in age from 15 to 53 years who underwent orthopedic surgery were studied at the Hospital of Hirosaki University School of Medicine. They belonged to ASA classifications 1 and 2. The orthopedic surgery included osteotomy of the femur, meniscectomy, removal of bone fumor in the extremities, arthroscopy, and others.

As premedication, patients received diazepam 10 mg and atropine 0.5 mg intramuscularly 1 h prior to the induction of anesthesia. Anesthesia was induced between 8:15 and 8:30 A.M., with incremental inhalation of sevoflurane from 1% to 5% combined with nitrous oxide (4 l/min) and oxygen (2 l/min). When an adequate depth of anesthesia was obtained, judged by loss of consciousness and central fixation of the pupil, inspired concentration of sevoflurane was decreased to 2%–3%. Patients were ventilated manually by mask for 20 min. Thereafter, succinylcholine 0.8–1 mg/kg was administered intravenously to facilitate tracheal intubation. The inspired concentration of sevoflurane was controlled to match the circulatory changes during the surgical procedures. Lactated Ringer's solution was infused intravenously during the procedure.

Endocrine parameters including plasma levels of cortisol, aldosterone, ACTH, growth hormone, epinephrine, and norepinephrine were measured. Systolic blood pressure, diastolic blood pressure, heart rate, serum electrolytes, and arterial blood gas analysis were also determined. Blood samples were collected or hemodynamic parameters measured at the following occasions: (a) just before the induction of anesthesia, which served as control; (b) 20 min after the induction of anesthesia; (c) immediately after tracheal intubation, (d) 30 min after the skin incision; (e) 60 min after the skin incision; and (f) after adequate recovery from anesthesia in the recovery room.

The Student's t test was used for the statistical analysis, and $P < 0.05$ was considered statistically significant. All figures are expressed as mean ± standard error (SE).

Results

Plasma cortisol level was unchanged during induction of sevoflurane anesthesia and the stimulation of tracheal intubation; it increased nonsignificantly with surgical stimulation 60 min after the skin incision. It reached a peak of 2.5 times the control level (significant difference) after the recovery from anesthesia (Table 1). Plasma aldosterone level was affected neither by sevoflurane anesthesia alone nor by the stimulation of tracheal intubation. Thereafter, it significantly increased gradually with surgical stimulation and reached a peak of 2.5 times the control value in the recovery room. Plasma ACTH level was unchanged with sevoflurane anesthesia alone, with tracheal intubation, and with surgical stimulation, but it increased significantly compared with the control level in the recovery room. The pattern of variation was similar to that of cortisol. Plasma growth hormone level was not affected by either sevoflurane anesthesia alone or by surgery. However, it increased significantly after the recovery from anesthesia. Plasma catecholamine levels (both epinephrine and norepinephrine) were unchanged with sevoflurane anesthesia alone, tracheal intubation, and surgical intervention.

Systolic blood pressure decreased significantly compared with preanesthetic levels during anesthesia and surgery, but this decrease was not clinically important. Blood pressure returned to the control level immediately after the recovery from anesthesia. Changes in diastolic blood pressure were similar to those in systolic blood pressure during anesthesia and surgery (Table 2). Heart rate was not changed with sevoflurane and surgery compared with the control value. Serum electrolytes and arterial blood gases remained within normal limits.

Discussion

Our study demonstrated that sevoflurane anesthesia itself did not affect the plasma levels of stress hormones of cortisol, aldosterone, ACTH, growth hormone, epinephrine, or norepinephrine. Plasma levels of these hormones have

Table 1. Effect of sevoflurane anesthesia on plasma levels of stress hormones during orthopedic surgery

Sampling time Plasma hormone levels	Preanesthesia	20 min after induction	Immediately after intubation	30 min after incision	60 min after incision	After recovery[a]
Cortisol (μg/dl)	12.0 ± 1.3	12.9 ± 1.9	10.3 ± 1.7	11.8 ± 2.0	17.7 ± 3.1	29.6 ± 1.4**
Aldosterone (ng/dl)	8.2 ± 0.4	10.2 ± 1.4	10.8 ± 1.6	13.5 ± 2.1*	15.4 ± 2.3**	20.4 ± 2.4**
ACTH (pg/ml)	23.5 ± 2.4	16.8 ± 2.4	36.2 ± 14.5	19.3 ± 4.5	65.1 ± 23.1	158.2 ± 39.0**
Growth hormone (ng/ml)	2.8 ± 0.2	3.3 ± 0.2	4.1 ± 0.7	5.2 ± 1.2	5.3 ± 1.9	6.9 ± 1.6*
Epinephrine (pg/ml)	28.2 ± 3.0	28.2 ± 6.1	45.4 ± 22.8	22.8 ± 4.3	28.0 ± 6.2	67.5 ± 36.1
Norepinephrine (pg/ml)	89.6 ± 19.7	130.8 ± 30.0	129.3 ± 29.2	88.8 ± 16.9	80.0 ± 19.1	74.2 ± 24.7

Each figure denotes mean ± SE.
* $p < 0.05$ versus preanesthesia value.
** $p < 0.01$ versus preanesthesia value.
[a] After adequate recovery from anesthesia in the recovery room.

Table 2. Effect of sevoflurane anesthesia on hemodynamic parameters during orthopedic surgery

	Preanesthesia	20 min after induction	Immediately after intubation	30 min after incision	60 min after incision	After recovery[a]
Inspired concentration of sevoflurane (%)	–	2.9 ± 0.1	2.9 ± 0.1	2.9 ± 0.2	2.8 ± 0.1	–
Systolic blood pressure (mmHg)	124 ± 2	109 ± 3**	110 ± 3**	108 ± 3**	112 ± 3**	123 ± 3
Diastolic blood pressure (mmHg)	73 ± 3	66 ± 3	63 ± 3*	60 ± 4*	64 ± 5	74 ± 4
Heart rate (beats/min)	84 ± 4	86 ± 3	94 ± 4	93 ± 3	94 ± 4	84 ± 3

Each figure denotes mean ± SE.
* $p < 0.05$ versus preanesthesia value.
** $p < 0.01$ versus preanesthesia value.
[a] After adequate recovery from anesthesia in the recovery room.

been reported to be unchanged with halothane or enflurane anesthesia (Oyama 1983).

We administered this new inhalation agent in orthopedic patients. Plasma levels of cortisol, ACTH, and growth hormone were not significantly changed during surgery but plasma levels of these hormones tended to increase gradually. They reached their highest levels after recovery from anesthesia (significant difference). Plasma levels of cortisol, aldosterone, ACTH, growth hormone have been reported to increase significantly during abdominal surgery and after recovery from anesthesia with halothane or enflurane anesthesia (Oyama 1983). The magnitude of surgical stress is thought to be smaller in orthopedic surgery than in abdominal intervention (T. Murakawa, unpublished, 1986), which might account for our inability to demonstrate a significant changes in plasma levels of these hormones due to surgery per se.

Only plasma aldosterone levels increased significantly during surgery and after the recovery from anesthesia in the present study. The reason why plasma aldosterone levels increased significantly during surgery in this study is unknown.

Plasma catecholamine levels were unchanged throughout the procedure. Sevoflurane may have the property of suppressing the sympathomimetic response to minor surgical stress. This was substantiated by the stable blood pressure and heart rate seen during sevoflurane anesthesia and surgery.

Advantages of sevoflurane include its noninflammable property, rapid anesthetic induction, fast emergence from anesthesia due to low blood gas partition coefficient, low metabolization in the body, acceptable odor, and stable circulation. Although sevoflurane has relatively less anesthetic potency and causes mild respiratory depression, significant adverse effects have not been found as yet.

Our study suggests that sevoflurane might be a useful inhalation anesthetic agents, but further evaluation is necessary.

Reference

Oyama T (1983) Endocrine response to general anesthesia and surgery. In: Oyama T (ed) Endocrinology and the anaesthetist. Elsevier, Amsterdam, pp 1–21 (Monographs in anaesthesiology, vol 11)

Magnetic Resonance Imaging of the Peridural Space

H. LAUBOECK and W. ISCHEBECK

Peridural anaesthesia has a number of advantages to the patient. For instance, cerebrospinal fluid (CSF) does not occur with this method. In addition the continuous catheter method permits precise regulation of the anaesthetic process itself as well as its maintenance and distribution. In contrast to general anaesthesia, peridural anaesthesia does not affect the life-preserving centres since it is restricted to the extradural nerve segments. Its therapeutic advantages are: increased vascular blood flow of the involved segments; active and passive mobilisation of joints; suitability in the treatment of pain by means of peridural opiate analgesia, for analgesia during labour, for surgical treatment, and when muscle relaxation is necessary. The patient can be awake or sedated, or a combination can be obtained with general anaesthia, and postoperative pain can be controlled.

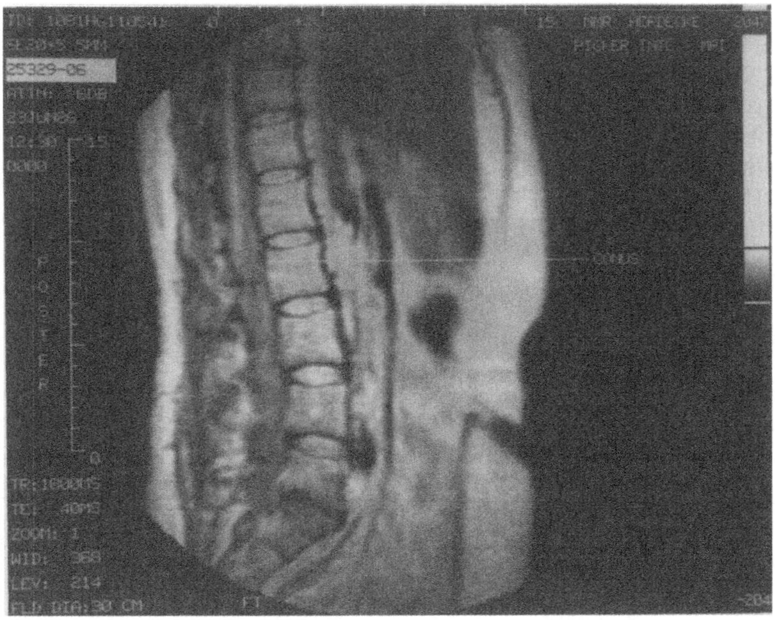

Fig.1.

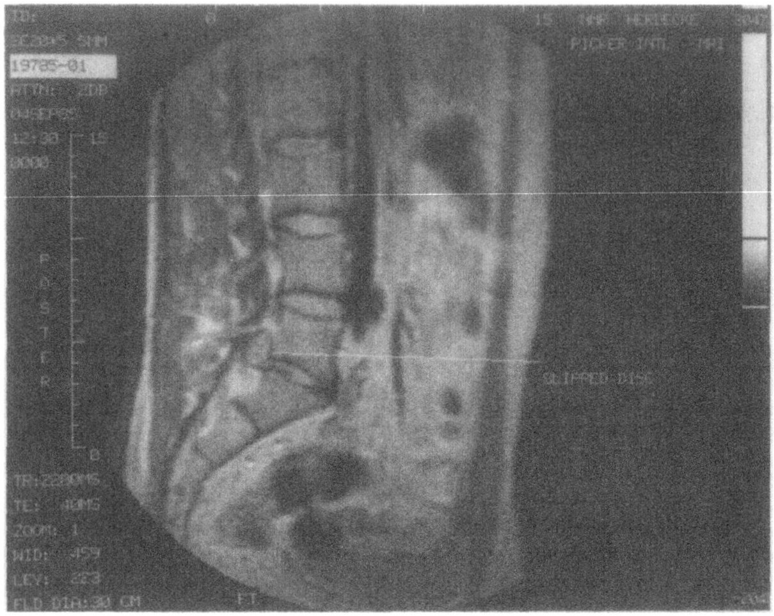

Fig. 2.

This method is also most advantageous regarding cardiocirculatory functions; due to vascular dilatation as a result of sympathetic nerve block, blood flow velocity is increased according to the Hagen-Poiseuille equation. In addition, an improvement in heart function occurs, as described by Laplace's, law: by reduction of myocardial wall tension, due to reduction of pre- and afterload and reduction of the heart rate, the demand for and consumption of oxygen is reduced. The cardiac index remains normal or can even be increased if the relative volume loss and the heart rate are compensated. The mechanism of artificial ventilation antagonistically impedes the venous return of blood into the thoracic cavity, depending on the magnitude of the mean airway pressure (Valsalva effect). The transmural myocardial pressure gradient is also decreased, which further decreases the diastolic filling of the heart. When people are breathing spontaneously, we know that the synergism between their own breathing movements and the central circulation supports the venous return of blood.

These features make it necessary to reinvestigate the structures of the spinal canal using the new method of imaging – nuclear magnetic resonance, or magnetic resonance (MR) tomography. This allows for a completely new way of imaging the spinal canal, in a manner which has never existed before. Since developments in anaesthesia have repeatedly shown how important precise knowledge of the anatomy of the spinal canal is to the success of anaesthesia, we investigated this anatomy and its subtle variations in various structures. This seems especially important since MR today plays a major role in the diagnosis of illness of the spine, and increasingly so in the case of spinal tumours.

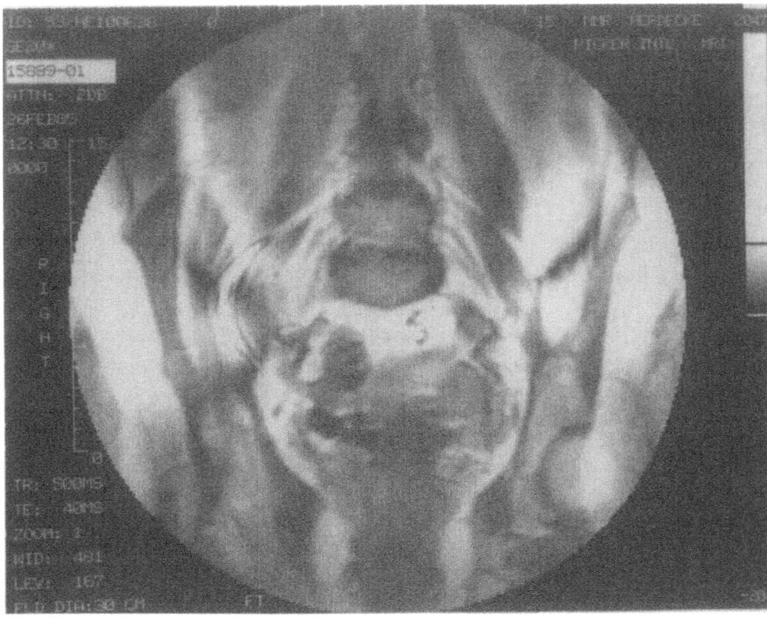

Fig. 3.

Such three-dimensional imaging in the medical field was unheard of prior to the invention of MR, a method which up to now has shown no side effects for the patient in any of the tomography instruments now in use. Our research was done with a Picker MR vista 2035 – 0,5 Tesla. This is not the occasion to explain the technique of MR in any detail, but we do wish to point out that in this case we are not dealing with an X-ray system which measures absorption but a method that visualizes various structures within the body by using different sequences and repetition times.

The conus can be easily differentiated, and its position in relation to the vertebra can be seen clearly. The spinal cord appears dark, the CSF having a slightly lighter signal intensity. It is of practical clinical importance to be able to establish the height of the conus, since variability is great from patient to patient, especially between men and women (Fig. 1).

Compared with the image in computed tomography (CT), the MR image has the advantage, by using the sagital plane, of enabling us to determine precisely the height and localisation of, for example, a slipped disc. This serves as an immediate indication of the degenerative process. One can clearly distinguish between the different signal intensities and the degree of protrusion of the nucleus pulposus (Fig. 2). In addition, a coronal image is shown (Fig. 3). Although these various ways of imaging show the structures differently, one must always remember that bone material has a lower signal response than structures containing more protons. In addition, Fig. 4 demonstrates images produced by using additional sequences, with a pronounced white signal of the CSF, similar to the case in X-ray myelography.

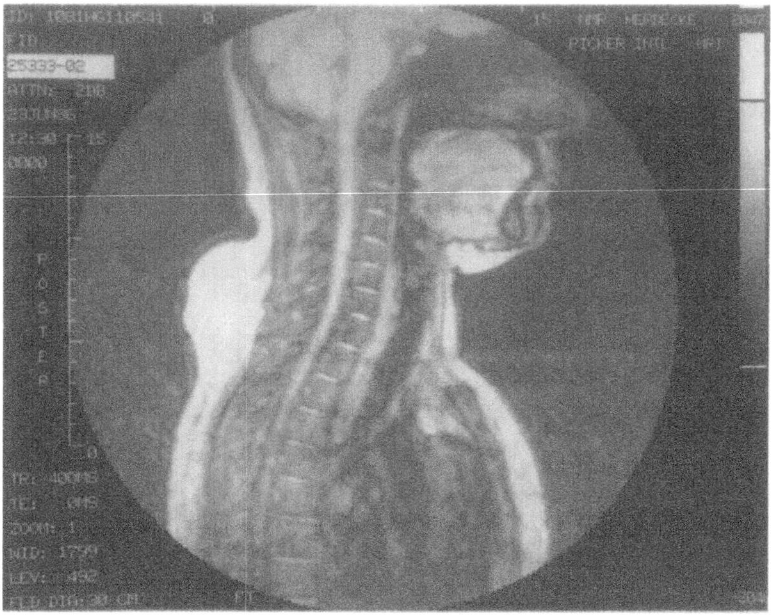
25333-02
PICKER INT. MRI
TR: 400MS
ZOOM: 1
WID: 1759
LEV: 492

Fig. 4.

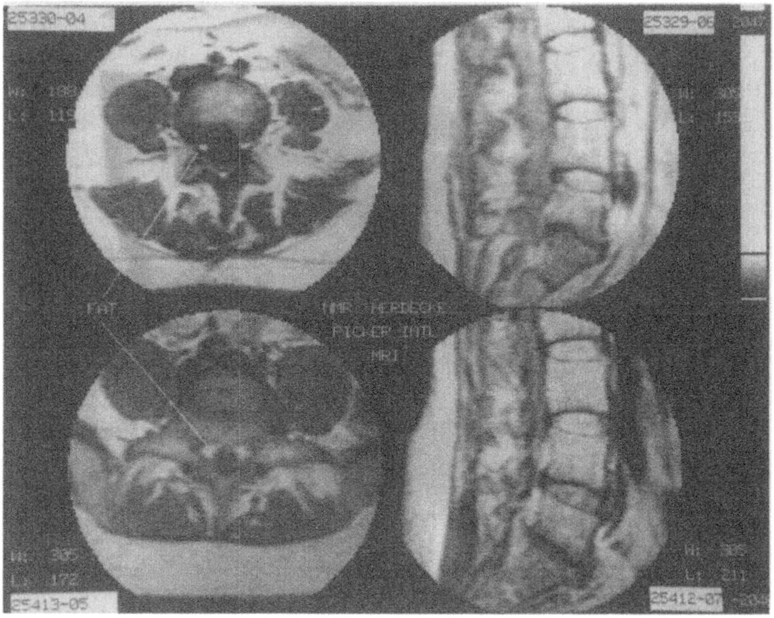
25330-04
25329-06
FAT
NMR HERDECKE
PICKER INT.
MRI
25413-05
25412-07

Fig. 5.

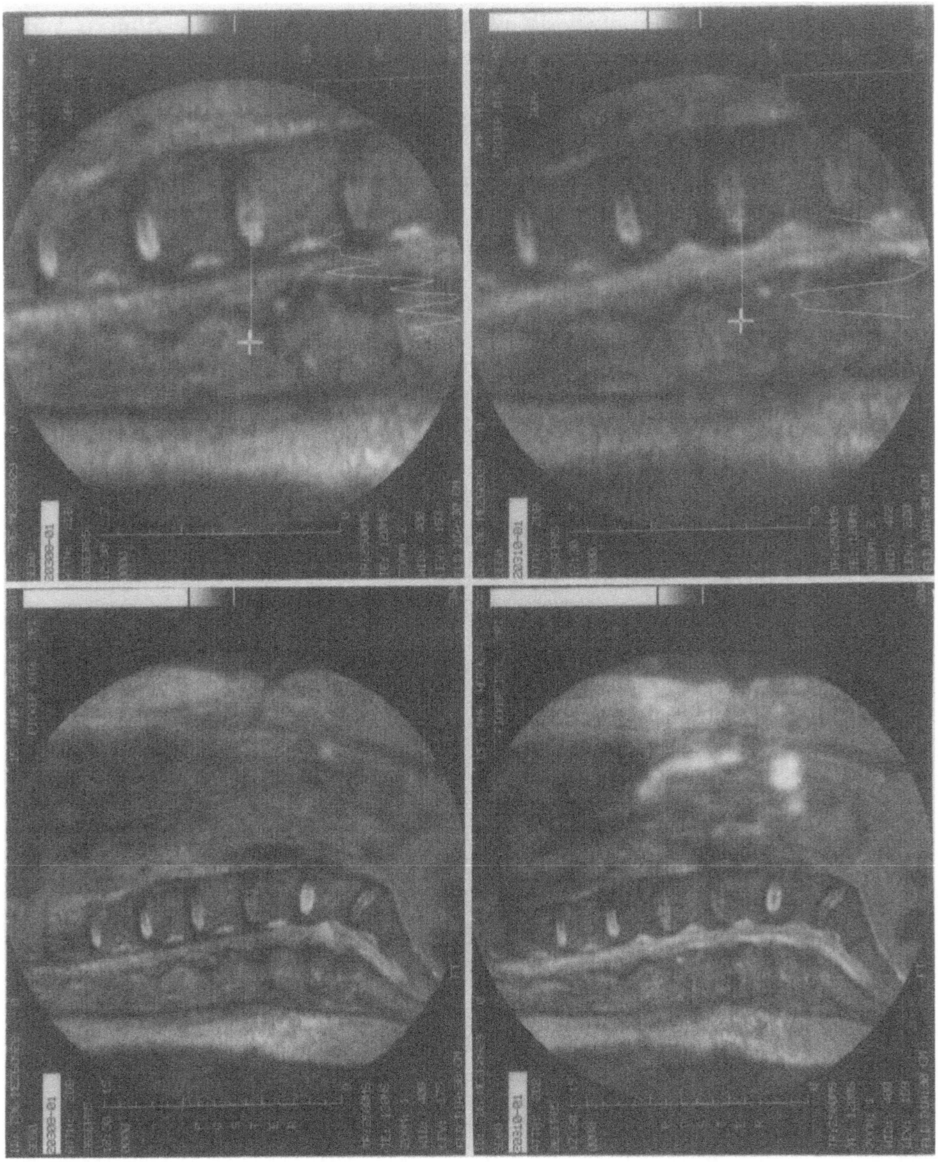

Fig. 6.

Due to the large variability of sequences and repetition times, MR imaging provides detailed information about the structures lying in the peridural and the intradural space. A low sequence demonstrates the impressive variation in the distribution of fat within the peridural space; the consequences that this may have for peridural anaethesia have not yet been established.

If one compares the sagittal section with transverse slides, it is possible to reach conclusions about the expansion of the peridural space. This is done by looking at the distribution of fat tissue, especially within the lumbar spinal canal but also in the upper regions. Systematic examination must yet establish the interaction between distribution of fat tissue in the peridural space and the beginning, intensity, distribution and duration of peridural anaethesia (Fig. 5).

Figure 6 presents the image of a patient who received an epidural injection (Bupivacain, Woelm) for therapeutic analgesia. We expected more signal resonance with the peridural space because of the locally increased concentration of water. Accordingly, we found clearly visible and measurable differences, the so-called profile, of the signal intensity.

The importance of peridural anaesthesia is increasing. Depending on the individual case, it sometimes seems justified, and for the safety of the patient necessary, to look at the anatomical structure of the spinal canal prior to administering peridural anaesthesia. MR imaging provides this opportunity without causing any risk to the patient. Although the expense of an MR examination hinders the use of such measures, we feel that more research should be carried out regarding their use in practical clinical medicine.

Hepatic Drug Metabolism in Multiple Organ Failure

W. SCHREGEL, H.-D. KUNTZ, and M. VITT

Introduction

Even today the mortality of patients with septic multiple organ failure is terribly high. Lazarus and Weis [17] have pointed out that most patients in intensive care units do not die of ARDS, but of septic complications. Patients with multiple organ failure die although ventilatory support and artificial renal replacement therapy have been given. Liver function and drug disposition in these patients have not been studied very intensively. On the other hand, the liver has a central position in metabolism of drugs, carbohydrates, and proteins [5]. Cases of severe over- and underdose of drugs have been reported in ICU patients [2, 8, 9]. The hepatic clearance of indocyanine green has been shown to correlate with survival in critically ill surgical patients [11]. There is some evidence that the liver plays an important role in recovery from cardiocirculatory shock and in prevention of septic complications [6], which are frequently associated with the highest postoperative bilirubin levels [5]. Little has changed since 1958, when Erskine [6] finished his review on "The relations of the liver to shock" with the words: "A great deal of investigative work remains to be done"!

Hepatic clearance of medicaments (C_e) is dependent on hepatic blood flow (Q) and hepatic extraction ratio (E):$C_e = Q \times E$. If the extraction ratio is above 60%, we can conclude from the equation that drug elimination is limited by hepatic blood flow. Drugs with a high extraction ratio are called flow-limited drugs. Examples of this type of drug disposition are lidocaine, β-blocking agents, morphine, and nitroglycerin [3] and in intensive care patients perhaps also hexobarbital and fentanyl, [25]. If the extraction ratio falls below 60%, drug disposition is limited by the drug-metabolizing capacity of liver cells [3]. Such capacity-limited drugs include benzodiazepines, barbiturates, theophylline, and most the other drugs used in anesthesia and intensive care medicine. Inside the liver two types of drug-metabolizing reactions are distinguishable: phase I reactions modify the drug molecule by demethylation, oxidation, or reduction [3] and phase II reactions involve the conjugation of the drug with glucuronic adic, sulfate, or glutathion and make excretion into the bile or urine possible [16].

Drug metabolism in critically ill patients can be influenced by various factors; surgical stress, barbiturates, or chronic treatment with antiepileptic drugs can induce hepatic microsomal enzymes [9]; drug levels are lower than in normal persons. During ICU treatment an increase in the number of liver cells and in the size of the liver is normally found [24]. After a prolonged shock a reduction in the number of liver cells can occur [6]. Hepatic blood flow is reduced by shock, heart failure, increased venous pressure, ventilation with PEEP, and by catecholamines and cimetidine [6, 7, 29]. Flow-limited drugs may reach toxic levels if the dose is not reduced. Reduced levels of proteins, especially albumin, or changes in protein binding decrease the amount of protein-bound drug [3]. The drug effects may be enhanced or reduced because a greater amount of nonbound drug can be eliminated by the excreting organs. The same effects can result from altered distribution volumes [23]. A carbohydrate-based diet results in a decreased activity of drug-metabolizing capacity, whereas a protein-based diet causes the opposite effect [22]. Cimetidine and antimycotic drugs have been reported to inhibit hepatic microsomal enzymes [7, 10]. Preexisting diseases of liver, heart and kidneys may favor increased or even decreased levels [29]. If renal failure exists, a reduced renal elimination of drugs, altered distribution volumes, and decreased protein binding must be considered [2]. During total liver failure drug metabolism is reduced to only 5% of the normal level, the hepatic protein synthesis fails and the function of the reticuloendothelial system may be severely depressed. Several authors agree that simple parameters for prediction of the activity of hepatic drug metabolism do not exist, and that "standard" liver tests are of limited value [3, 9, 11, 23].

For patients with multiple organ failure the situation is even more complex. Sepsis is the etiology in most cases. Surgical treatment of the location of infection should be tried whenever possible; antibiotic and antimycotic drugs are applied regularly. Furthermore, respirator therapy, hemodialyses or hemofiltration, Swan-Ganz catheterization, and invasive blood pressure monitoring are necessary. Protein levels are low due to capillary leaks, and an increased turnover rate results, from decreased synthesis [4]. Coagulation disorders and gastric bleeding may occur. Many patients need blood and substitution of plasma proteins. Patients with multiple organ failure are dependent on a great range of drugs, e.g., catecholamines, digitalis, nitroglycerin, heparin, insulin, antacids, and perhaps corticosteroids.

Most recommendations for drug therapy in renal failure are based on an unaltered liver function. On the other hand, general plasma protein depletion [26], increased levels of bilirubin and liver enzymes, and liver dysfunction have been reported in septic intensive care patients [8, 20]. Moreover, humoral and cellular defense mechanisms may be depressed [1, 21]. Fibronectin is a glycoprotein produced by liver cells, and it enables the reticuloendothelial system to remove bacteria, toxins, and cell fragments from the circulation [12]. A reduction in fibronectin has been found after severe trauma, burn injury, and in patients with multiple organ failure of septic etiology [12]. In few cases their condition improved significantly after substitution of fibronectin [27], whereas a lower, but not statistically significant mortality was found in a larger investigation [19].

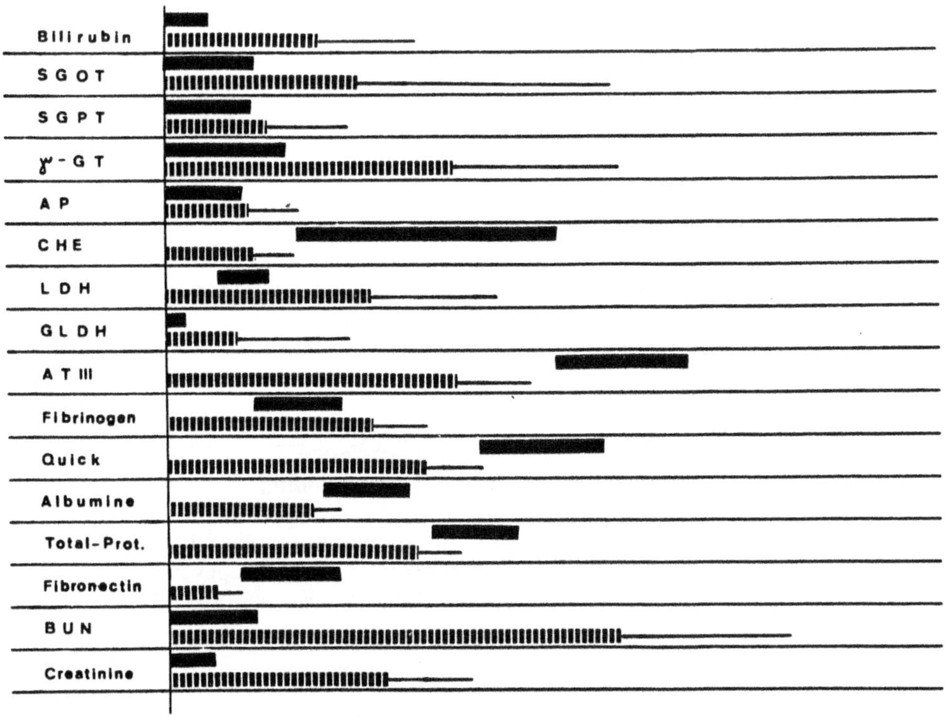

Fig. 1. Liver values and functionally important proteins. *Black bars,* normal range; *striped bars,* observed values. *SGOT,* Serum glutamic oxolacetic transaminase; *SGPT,* serum glutamic pyruvic transaminase; *γ-GT,* gamma-glutamyltransferase *AP,* Alkaline Phosphatase; *CHE,* Cholinesterase *LDH,* lactic dehydrogenase; *GLDH,* glutamic dehydrogenase. *AT III,* antithrombin III; *BUN,* blood urea nitrogen

Therefore, we conducted a clinical investigation to answer the following questions:

- Is the liver one of the failing organs during multiple organ failure?
- Which functions of hepatic drug metabolism are influenced?
- Are there any correlations to "standard" liver tests?
- Which recommendations can be given for drug therapy during multiple organ failure?
- What are the levels of functionally important proteins during multiple organ failure?

Patients and Methods

The experimental design of the study was presented to the Ethical Committee of the Ruhr University in Bochum (FRG). In previous investigations patients with hepatic, cardiac, and renal diseases have been investigated with these methods without complications. Our study included 15 patients with respiratory and renal failure. The investigation started when artificial renal replacement

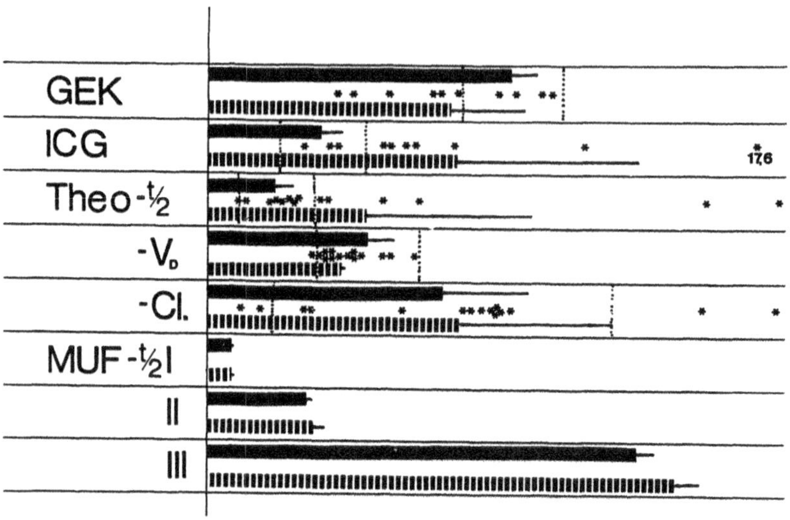

Fig. 2. Partial functions of drug metabolism. *Asterisks* indicate individual values. *GEK,* Galactose elimination capacity; *ICG,* indocyanine green; *Theo,* theophylline; *t/2,* half-life; V_o, distribution volume; *Cl.,* clearance; *MUF,* 4-methylumbelliferon, (t/2 I, II, III according to the triphasic elimination of the drug

therapy had to be applied. In six cases the tests were performed on patients with continuous hemofiltration, in the other nine cases shortly before or after intermittent hemodialysis or hemofiltration was performed. Blood gases were normalized, and systolic blood pressure was maintained at a level higher than 100 mmHg. Patients with preexistent liver disease or theophylline medication were excluded.

Bilirubin, liver enzymes, and the levels of important proteins were evaluated in all patients. Galactose elimination capacity reflects the "functional hepatocyte mass" and served for classification of severity of various liver diseases [14, 15]. Indocyanine green has a high extraction ratio and can be taken as an estimate of flow-limited hepatic elimination [15]. Theophylline metabolism occurs in the liver by demethylation and/or oxidation [13]; therefore, it can serve for estimation of hepatic microsomal enzyme activity (phase I). 4-Methylumbelliferon (Hymecromon) is mainly excreted into the bile after glucuronidation by the liver and is an indicator of the glucuronidation capacity of the liver (phase II) [15].

Results and Discussion

In accordance with reports in literature [18, 20], elevations of bilirubin and liver enzymes were also found in our investigation (Fig. 1). These values do not indicate liver cell necrosis, which is usually correlated with high levels of SGOT and SGPT. The high level of lactic dehydrogenase (LDH) may indicate hemo-

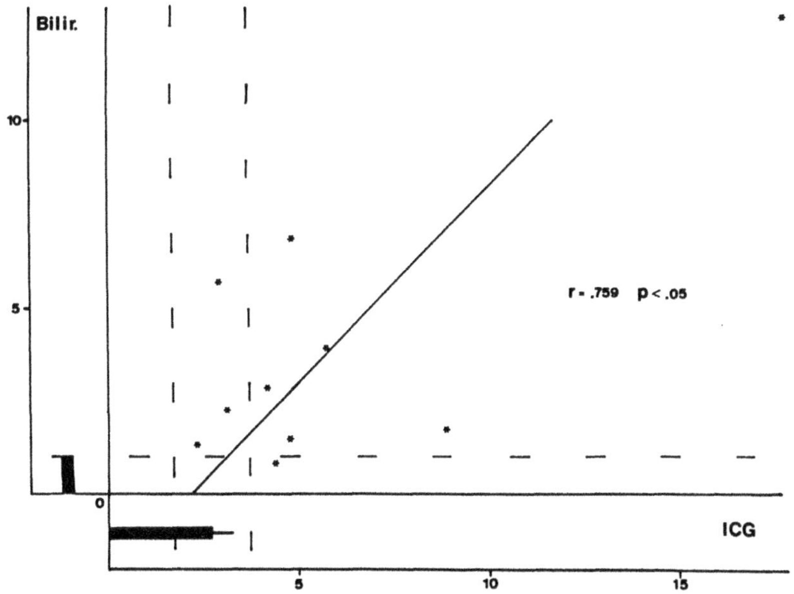

Fig. 3. Correlation (r = 0,759) between bilirubin and indocyanine green *(ICG)*

lysis, whereas jaundice in septic patients is thought to be a result of intrahepatic cholestasis [5, 18]. The levels of functionally important proteins are lowered in nearly all patients, although most received great amounts of fresh-frozen plasma and albumin. Antithrombin III plays an important role in prevention of intravascular coagulation, which has been seen as responsible for deterioration in pulmonary and renal function [28]. Fibronectin levels are also decreased, indicating a severe depression in reticuloendothelial system function [12]. These results may help to explain the high mortality of patients with multiple organ failure, although their location of infection and their pulmonary, renal, and cardiovascular problems were apparently treated sufficiently. The normal detoxifying role of the liver may simply be overwhelmed, and endotoxin may exert its toxic effects on coagulation, microcirculation, kidney, and heart [18].

Figure 2 demonstrates the results of drug metabolism. Glucuronidation of drugs, measured by the half-life of 4-methylumbelliferon, is not influenced in any patient. Unfortunately, only few drugs are eliminated by glucuronidation; oxazepam is not available for intravenous application, and morphine is also dependent on renal elimination [2]. The other three tests showed abnormal results in some patients. Galactose elimination capacity is reduced to deleterious levels in patients with hepatic cirrhosis [14]. Indocyanine green elimination was found to be pathological in six out of ten patients. Theophylline distribution volume was not significantly different from normals. Thus, the increased half-life levels reflect the reduced function of hepatic microsomal drug metabolism [13].

Correlations have been calculated between liver enzyme values, protein levels, and parameters of drug metabolism. The results are depicted in the

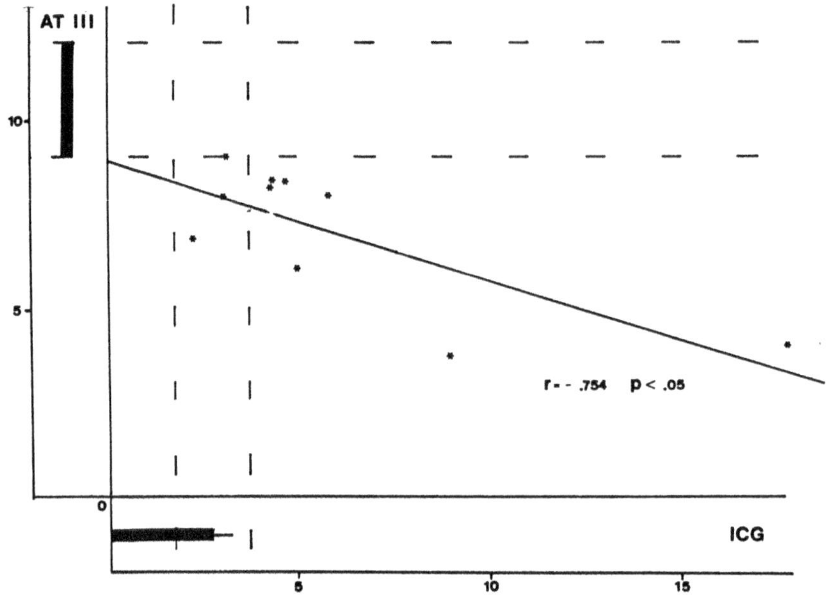

Fig. 4. Correlation (r = 0,754) between antithrombin III *(AT III)* and indocyanine green *(ICG)*

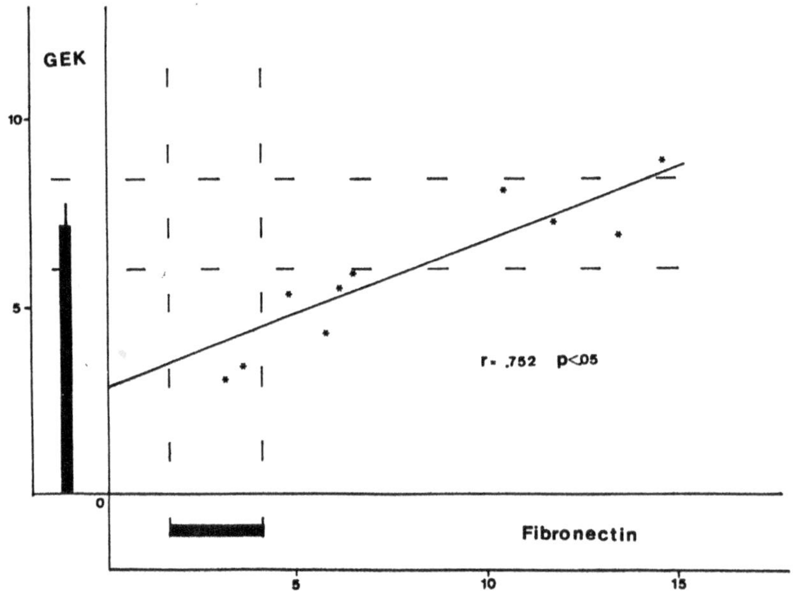

Fig. 5. Correlation (r = 0,752) between galactose elimination capacity *(GEK)* and fibronectin

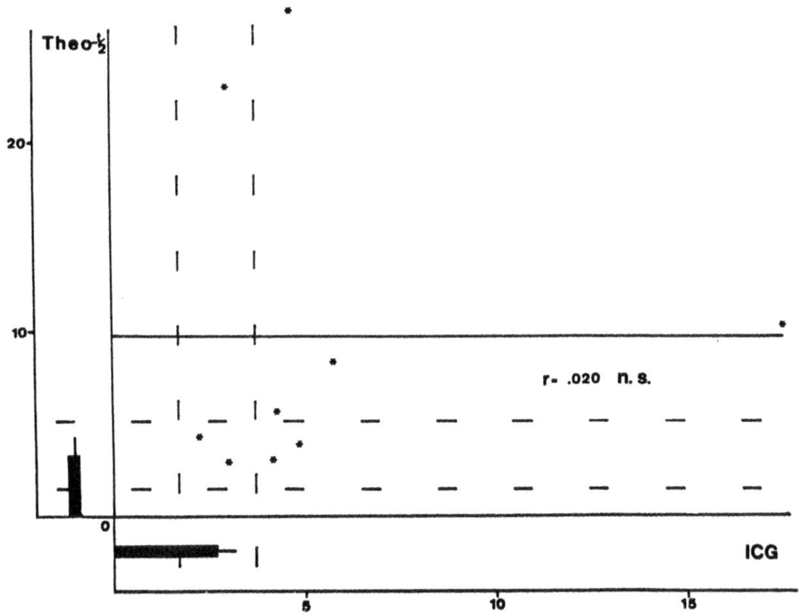

Fig. 6. Relationship between theophylline half-life *(Theo-t/2)* and indocyanine green *(ICG)*

remaining figures. Only three significant correlations exist: between indocyanine green and bilirubin (Fig. 3), between indocyanine green and antithrombin III (Fig. 4), and between fibronectin and galactose elimination capacity (Fig. 5). Perhaps bilirubin, antithrombin III, and fibronectin reflect hepatic partial functions, but in spite of the significant correlations it does not seems justified to conclude about drug metabolism in the individual patient. This hypothesis gains additional support by analysis of the individual values. Although there is a significant correlation between bilirubin and indocyanine green (Fig. 3), we can find at nearly identical bilirubin levels normal, slightly, and extremely prolonged levels of indocyanine green half-life. Figures 6 and 7 demonstrate that no correlation exists between the drug-metabolizing functions. Of course, more investigation is necessary to identify patient groups at high risk from alterations in drug metabolism [9].

Conclusions

We may conclude from the results of this investigation the following:

- Total hepatocellular necrosis does not occur frequently during multiple organ failure. The functional hepatocyte mass may be reduced in some cases.
- In some patients partial functions of hepatic drug metabolism may be influenced severely and independently of each other. Glucuronidation was not impaired in any case. Toxic drug levels as well as drug underdose may occur with the same patient at the same point of time.

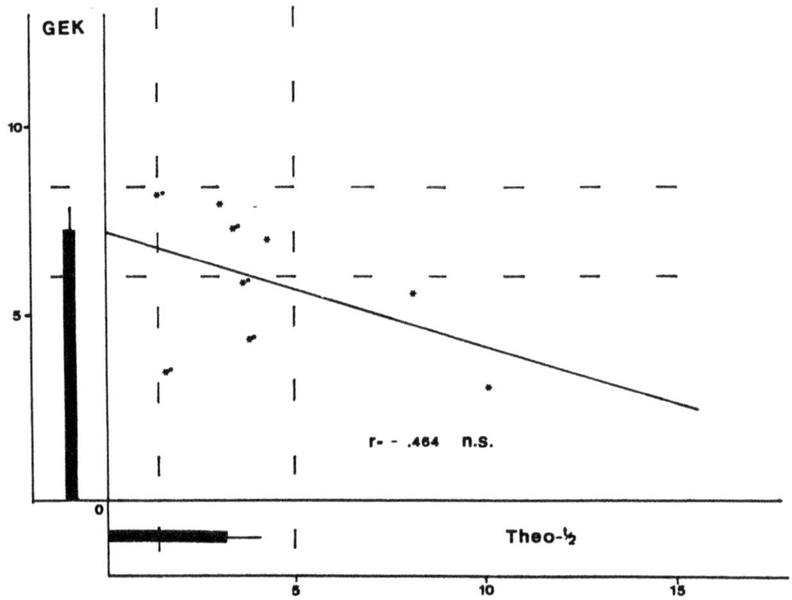

Fig. 7. Relationship between galactose elimination capacity *(GEK)* and theophylline half-life *(Theo-t/2)*

- At the present state of research, drug level monitoring [23] and a careful clinical observation are necessary to avoid drug over- or underdose.
- Hepatic protein synthesis is not sufficient to normalize the levels of functionally important proteins such as albumin, antithrombin III, or fibronectin.
- Multiple organ failure will remain a challenge in the future, and progress will be made only step by step. Prevention of respiratory or renal failure, earlier application of hemofiltration, antithrombin III, or fibronectin and a more effective drug therapy may be some of these steps.

References

1. Abraham E, Chang Y (1986) Cellular and humoral bases of hemorrhage-induced depression of lymphocyte function. Crit Care Med 14:81–86
2. Ball M, McQuay HJ, Moore RA, Allen MC, Fisher A, Sear J (1985) Renal failure and the use of morphine in intensive care. Lancet 1:784–786
3. Blaschke TF (1977) Protein binding and kinetics of drugs in liver diseases. Clin Pharmacokin 2:32–44
4. Caldwell TB (1981) Infections diseases. In: Katz J, Benumof J, Kadis LB (eds) Anaesthesia and uncommon diseases, 2nd edn. Saunders, Philadelphia, pp 588–634
5. Champion HR, Jones RT, Trump BF, Decker R, Wilson S, Stega M, Nolan J, et al (1976) Post-traumatic hepatic dysfunction as a major etiology in post-traumatic jaundice. J Trauma 16:650–657
6. Erskine JM (1958) The relations of the liver to shock. Int Abstr Surg 106:207–214
7. Feely J, Wilkinson GR, Wood AJJ (1981) Reduction of liver blood flow and propanolol metabolism by cimetidine. N Engl J Med 304:692–695

8. Fuchs KH, Maier C, Engelke M, Wirtz HJ (1984) Untersuchungen zur Pharmakokinetik von Cimetidin bei kontinuierlicher arteriovenöser Hämofiltration. Anaesthesist [Suppl 457]:53
9. Gramm HJ, Heinemeyer G, Dennhardt R, Roots J (1985) Veränderungen des Arzneimittelmetabolismus bei Intensivpatienten. In: Rügheimer E, Pasch T (eds) Notwendiges und nützliches Messen in Anaesthesie und Intensivmedizin. Springer, Berlin Heidelberg New York, pp 496–499
10. Heinemeyer G, Roots J, Schulz H, Dennhardt R (1985) Hemmung der Pentobarbital-Elimination durch Miconazol bei Intensivtherapie des erhöhten intracraniellen Drucks. Intensivmedizin 22:164–167
11. Kholoussy AM, Pollack D, Matsumoto T (1984) Prognostic significance of indocyanine green in critically ill surgical patients. Crit Care Med 12:115–116
12. Klingemann HG (1986) Fibronectin – Indikator bei Zelluntergang. Diagn Lab 36:1–7
13. Kuntz HD, May B (1982) Theophyllin – Elimination bei chronischen Lebererkrankungen. Med Welt 33:1710–1711
14. Kuntz HD, May B (1983) Die Galaktose-Eliminationskapazität. Med Welt 34:646
15. Kuntz HD, May B (1983) Quantitative Leberfunktionsdiagnostik. MMW 125:678–680
16. Kuntz HD, Femfert U, May B (1983) Pharmakokinetik von 7-Hydroxy-4-Methyl-2-H-benzofuran-2-on als Funktionsprüfung für die Glukuronidierungskapazität der Leber. Verh Dtsch Ges Inn Med 89:630–633
17. Lazarus G, Weis KH (1985) Extrakorporale CO_2-Elimination beim schweren ARDS als randomisierte Studie? Anaesth Intensivmed 26:311–313
18. Ledingham J (1985) Hepatocellular dysfunction in the critically ill. In: Vincent JL (ed) Update in intensive care and emergency medicine. Springer, Berlin Heidelberg New York, pp 56–59
19. Lundsgaard-Hansen P, Rubli E, Dorau JE, Papp E, Morgenthaler JJ (1985) A controlled trial of purified fibronectin in patients with severe abdominal infections. Anaesthesist [Suppl 95] 34
20. Melzer A, Haldemann G, Reist K (1984) Einfluß von Nutriflex 48 auf das Plasmaaminosäurenmuster chirurgischer Intensivpatienten mit Sepsis. Infusionstherapie 2:190–194
21. Nishijima MK, Takezawa J, Hosotsubo KK, Takahashi H, Shimada Y, Yoshiya J (1986) Serial changes in cellular immunity of septic patients with multiple-organ-system failure. Crit Care Med 14:87–91
22. Pantuck EJ, Pantuck CB, Weissman C, Askanazi J, Conney AH (1984) Effects of parenteral regimens on oxidative drug metabolism. Anesthesiology 60:534–536
23. Rietbrock J (1977) Beeinflussung der Arzneimittel-Elimination bei Patienten auf einer Beatmungsstation. Anasth Inform 18:58–70
24. Rietbrock J, Lazarus G (1979) Leberfunktion unter Anästhesie und Intensivbedingungen. Med Klin 74:1790–1794
25. Rietbrock J, Lazarus G, Richter E, Breimer DD (1981) Hexobarbital disposition at different stages of intensive care treatment. Br J Anaesth 53:283–293
26. Rubli E, Buessard S, Frei E, Lundsgaard-Hansen P, Pappova E (1983) Plasma fibronectin and associated variables in surgical intensive care patients. Ann Surg 197:310–317
27. Saba TM, Blumenstock FA, Scovill WA, Bernhard H (1978) Cryoprecipitate reversal of opsonic alpha-2-surface binding glycoprotein deficiency in septic surgical and trauma patients. Science 201:622–626
28. Thaler E, Kleinberger G (1977) Sepsis und Blutgerinnung. Intensivmedizin 16:54–59
29. Thomson PT, Rowland M, Melman KL (1971) The influence of heart failure, liver disease and renal failure on the disposition of lidocain in man. Am Heart J 82:117–122

GPSR Compliance

The European Union's (EU) General Product Safety Regulation (GPSR) is a set of rules that requires consumer products to be safe and our obligations to ensure this.

If you have any concerns about our products, you can contact us on ProductSafety@springernature.com

In case Publisher is established outside the EU, the EU authorized representative is:

Springer Nature Customer Service Center GmbH
Europaplatz 3
69115 Heidelberg, Germany

Batch number: 10704771

Printed by Printforce, the Netherlands